A Volume in The Laboratory Animal Pocket Reference Series

The Laboratory

RABBIT

The Laboratory Animal Pocket Reference Series

Published and Forthcoming Titles

The Laboratory Rabbit
The Laboratory Non-Human Primates
The Laboratory Mouse
The Laboratory Guinea Pig
The Laboratory Rat
The Laboratory Hamster and Gerbil
The Laboratory Cat
The Laboratory Small Ruminant

A Volume in The Laboratory Animal Pocket Reference Series

The Laboratory
RABBIT

Mark A. Suckow, D.V.M.
Diplomate ACLAM

Fred A. Douglas, B.S., LATG

CRC Press
Boca Raton   New York

Illustrations by Robert T. Burke

Acquiring Editor: Marsha Baker
Editorial Assistant: Jean Jarboe
Project Editor: Helen Linna
Marketing Manager: Susie Carlisle
Direct Marketing Manager: Becky McEldowney
Cover design: Denise Craig
PrePress: Kevin Luong
Manufacturing: Sheri Schwartz

Library of Congress Cataloging-in-Publication Data

Suckow, Mark A.
    The laboratory rabbit/ Mark A. Suckow and Fred A. Douglas.
        p.   cm. — (The laboratory animal pocket reference series)
    Includes bibliographical references and index.
    ISBN 0-8493-2561-7
    1. Rabbits—research. 2. Biology— laboratory animals. I. Suckow,
Mark A. II. Title. III. Series.
    BR749.H79G87 1996
    616′.0149—dc20
                                                        96-25117
                                                        CIP

No claim to original U.S. Government works
International Standard Book Number 0-8493-2561-7
Library of Congress Card Number 96-25117
Printed in the United States of America 1 2 3 4 5 6 7 8 9 0
Printed on acid-free paper

dedication

M.A.S.: My role in this book would not be possible were it not for the influence of several individuals: my parents, Gilbert and Gertrude, for their sense of perseverance; my children, Brianne and Abby, for their sense of belief; and my wife, Susan, for her unflagging love, patience, and encouragement.

F.A.D.: For my part in this book, thanks go to several colleagues: to Dr. Mark Suckow, a friend and mentor in the field, for his constant support and inspiration; to a dear departed colleague, Dr. Terry Reed, whose love for the domestic rabbit has been a sense of encouragement to me to share the knowledge with all who will listen; and last, but not least, to my friend and helpmate, Penny, whose interest and support in my career have enriched my life many times over.

preface

The use of laboratory animals, including rabbits, continues to be an important part of biomedical research. In many instances, individuals performing such research are charged with broad responsibilities, including animal facility management, animal husbandry, regulatory compliance, and performance of technical procedures directly related to research projects. In this regard, this handbook was written to provide a quick reference source for investigators, technicians, and animal caretakers charged with the care and/or use of rabbits in a research setting. It should be particularly valuable to those at small institutions or facilities lacking a large, well-organized animal resource unit and to those individuals who need to conduct research programs using rabbits starting from scratch.

This handbook is organized into six chapters: "Important Biological Features" (Chapter 1), "Husbandry" (Chapter 2), "Management" (Chapter 3), "Veterinary Care" (Chapter 4), "Experimental Methodology" (Chapter 5), and "Resources" (Chapter 6). Basic information and common procedures are presented in detail. Other information regarding alternative techniques or details of procedures and methods which are beyond the scope of this handbook is referenced extensively so that the user is directed toward additional information without having to wade through a burdensome volume of detail here. In this sense, this handbook should be viewed as a basic reference source and not as an exhaustive review of the biology and use of the rabbit.

The final chapter, "Resources," provides the user with lists of possible sources and suppliers of additional information, rabbits, feed, sanitation supplies, cages, and research and veterinary supplies. The lists are not exhaustive and do not imply endorsement of listed suppliers over suppliers not listed. Rather, these lists are meant as a starting point for users to develop their own lists of preferred vendors of such items. Table 14 lists vendors of cages and research and veterinary supplies by number and is followed by a list of contact information for these suppliers.

A final point to be considered is that all individuals performing procedures described in this handbook should be properly trained. The humane care and use of rabbits is improved by initial and continuing education of personnel and will facilitate the overall success of programs using rabbits in research, teaching, or testing.

authors

Mark A. Suckow, D.V.M., is Assistant Director of the Laboratory Animal Program at Purdue University in West Lafayette, IN.

Dr. Suckow earned the degree of Doctor of Veterinary Medicine from the University of Wisconsin in 1987, and completed a post-doctoral residency program in laboratory animal medicine at the University of Michigan in 1990. He is a Diplomate of the American College of Laboratory Animal Medicine.

Dr. Suckow has published over 30 scientific papers and chapters in books. His primary research interest is in mucosal immunology, with particular reference to *P. multocida* infection in rabbits.

Dr. Suchow was honored as the 1996 Young Investigator of the Year by the American Association of Laboratory Animal Science.

Fred A. Douglas is the Manager of the Veterinary Laboratory Animal Care unit and an instructor in the Veterinary Technology Program at the School of Veterinary Medicine at Purdue University in West Lafayette, IN.

Mr. Douglas earned the Bachelor of Science degree in Animal Science from the University of Nebraska in 1974, and graduated from the AALAS Institute for Laboratory Animal Management in 1994. He is a certified Laboratory Animal Technologist.

Mr. Douglas joined the Purdue University School of Veterinary Medicine animal care staff in 1982, and was appointed Manager shortly thereafter. In that capacity, he oversees the daily operation of the research animal unit and has teaching and regulatory compliance responsibilities.

Mr. Douglas has held a variety of elected and appointed positions in the American Association for Laboratory Animal Science and the Laboratory Animal Management Association. He has numerous presentations and publications on topics including animal care personnel management, animal technician training, and rabbit care and management to his credit.

Mr. Douglas was awarded the 1996 Charles River Medallion for excellence in laboratory animal management.

contents

important biological features

introduction

The rabbit has been and continues to be used extensively in research and testing. The domestic rabbit belongs to the taxonomic order Lagomorpha. Although the rabbit appears rodent-like with respect to some morphologic features, protein sequence data suggest that rabbits may be more closely related to primates than to rodents.[1] The genus/species name of the rabbit is *Oryctolagus cuniculus*. An adult male rabbit is called a **buck**, an adult female rabbit is called a **doe**, and an immature rabbit is called a **kit**.

Common uses of rabbits in research include polyclonal antibody production, biomedical device testing and development, and pyrogen and teratogenicity testing of pharmaceutical compounds.

breeds

The most common breed used in research is the albino New Zealand White (Figure 1). Less commonly used breeds include the Dutch Belted and the Flemish Giant.

Fig. 1.   An adult New Zealand White rabbit.

behavior

Rabbits are generally timid and nonaggressive, although an occasional animal will display aggressive defense behavior, including thumping of the cage floor with the rear feet, charging the front of the cage, and biting. Adult males frequently are more aggressive than females. Overly aggressive animals may need to be housed individually. Adults often urinate and defecate habitually in specific locations of the cage. Although the rabbit is naturally nocturnal, laboratory-housed rabbits exhibit diurnal patterns of behavior.[2]

anatomic and physiologic features

Important and unique anatomic and physiologic features of the rabbit include:

Dentition

- The dental formula of rabbits is 2(2/1 incisors, 0/0 canines, 3/2 premolars, and 2–3/3 molars).

- The teeth are all continuously erupting and will overgrow if malocculsion occurs.

- A set of small incisors directly behind the front incisors is referred to as the **"peg" teeth**.

Fragile Skeleton

- Comprises only 7% of the total body weight. In contrast, the skeleton of the cat contributes 13% to the total body weight in that species.[3]

- Predisposes rabbits to fracture of the lumbar spine if not restrained properly.

Cutaneous Structures

- The **ears** are long and slender, highly vascular, and serve both thermoregulatory and auditory functions. Rabbits should NOT be grasped nor restrained by the ears, since the ears are sensitive and easily damaged.

- The **dewlap** is a large fold of skin present on the neck of some rabbits. This area frequently accumulates moisture and is predisposed to dermatitis.

- Bucks have no nipples, while does have 8 or 10.

Gastrointestinal System

- The **cecum** is very large.

- The **sacculus rotundus** is a pale-appearing accumulation of lymphoid tissue located at the junction of the cecum and the ileum.

- **Peyer's patches**, pale accumulations of lymphoid tissue located along the ileum, are relatively large in rabbits compared to other species.

- The **cecal appendix** is a large accumulation of lymphoid tissue located at the distal end of the cecum.

- Rabbits normally produce both dry and moist forms of feces. During the daytime, dry feces composed largely of undigestable fiber is produced, while at night soft, moist **night feces** is produced. Night feces is largely a product of cecal fermentation, is an important source of B-complex vitamins and protein, and is normally ingested by the rabbit.

Urogenital System

- The **inguinal canals**, connecting the abdominal cavity to the inguinal pouches, do not close in the rabbit. For this reason, the superficial inguinal ring should be closed following orchiectomy by open technique, to prevent herniation.[4]

- **Inguinal pouches** are located lateral to the genitalia in both sexes. The pouches are blind and contain scent glands that produce white to brown secretions which may accumulate in the pouch.

- The **urethral orifice** of the buck is rounded, whereas that of the doe is slit-like. This feature is useful for distinguishing the sexes.

- The reproductive tract of the doe is characterized by two uterine horns which are connected to the vagina by separate cervices.

- The placenta is hemochorial, in which maternal blood flows into sinus-like spaces where the transfer of nutrients to fetal circulation occurs.[5] Passive immunity is transfered to the young via the yolk sac prior to birth rather than by the colostrum or the placenta.

- The urine is normally cloudy due to a large amount of calcium carbonate and ammonium magnesium phosphate (triple phosphate) crystals. Normal color may range from yellow to red or red brown. The pH of rabbit urine is typically alkaline.

normative values

Basic Biologic Parameters

Typical values for basic biologic parameters (Table 1), clinical chemistry (Table 2), cerebrospinal fluid (Table 3), interstitial fluid (Table 4) respiratory and cardiovascular function (Table 5), hematology (Table 6), and reproduction (Table 7) are presented below.

> **Note:** Values shown are representative of those in New Zealand White rabbits. Significant variation of values may occur between individual rabbits, breeds, and laboratories. It is imperative that individual laboratories establish normal values for their specific facility.

TABLE 1. BASIC BIOLOGIC PARAMETERS OF THE RABBIT

Parameter	Typical value	Reference(s)
Diploid chromosome number	44	6
Life span (years)	5–7	7
Number of mammary glands	8 or 10	3
Body weight (kg)	2–5	7
Body temperature (°C)	38–40	7
Metabolic rate (kcal/kg/d)	44–54	8
Food intake (g/kg/d)	50	7
Water intake (ml/kg/d)	50–100	7
Fecal production (g/d)	15–60	9
GI transit time (h)	4–5	3
Urine volume (ml/kg/d)	50–75	7
Urine specific gravity	1.003–1.036	7
Urine pH	8.2	7

Clinical Chemistry

Approximate values for clinical chemistry parameters are shown in Table 2. Values can be expected to vary between individual rabbits, breeds, laboratories, and with sampling method. It is imperative that laboratories establish normal values for their specific facility.

TABLE 2.   CLINICAL CHEMISTRY VALUES OF THE RABBIT

Parameter	Typical value	Reference(s)
Total protein (g/dl)	5.0–7.5	7,10–13
Globulin (g/dl)	1.5–2.7	7
Albumin (g/dl)	2.7–5.0	7,10,11,13
Acid phosphatase (IU/l)	0.3–2.7	7
Alkaline phosphatase (IU/l)	10–86	7,12
LDH (IU/l)	33.5–129	7,13
gamma-Glutamyl transferase (IU/l)	10–98	7
Aspartate aminotransferase (IU/l)	20–120	7,12
Creatine kinase (IU/l)	25–120	7
ALT (SGPT) (IU/l)	25–65	7,12,13
Sorbitol dehydrogenase (U)	170–177	13
Blood urea nitrogen (mg/dl)	5–25	7,10,12,13
Creatinine (mg/dl)	0.5–2.6	7,10,12,13
Glucose (mg/dl)	74–148	7,10,12,13
Sodium (meq/l)	125–150	7,10,12,13
Chloride (meq/l)	92–120	7,10,12,13
Potassium (meq/l)	3.5–7.0	7,10,12,13
Phosphorus (mg/dl)	4.0–6.0	7,10,12
Total bilirubin (mg/dl)	0.2–0.5	7
Uric acid (mg/dl)	1.0–4.3	7
Amylase (IU/l)	200–500	7
Serum lipids (mg/dl)	150–400	7
Phospholipids (mg/dl)	40–140	7,13
Triglycerides (mg/dl)	50–200	7,12
Cholesterol (mg/dl)	10–100	7,10,12,13
Corticosterone (µg/dl)	1.54	13

TABLE 3.   CEREBROSPINAL FLUID CELLULAR VALUES OF THE RABBIT

Parameter	Typical value	Reference(s)
White blood cells (cells/mm^3)	0–7	14
Lymphocytes (%)	40–79	14
Monocytes (%)	21–60	14

Cerebrospinal Fluid

Procedures for sampling cerebrospinal fluid are described in Chapter 5. Only very few white blood cells are present in normal cerebrospinal fluid. Presence of red blood cells in the sample often indicates contamination by blood during the sampling procedure.

TABLE 4. VALUES FOR INTERSTITIAL FLUID OF THE RABBIT

Parameter	Typical value	Reference
Colloid osmotic pressure (mmHg)	13.6	15
Viscosity (relative to water = 1)	1.9	15
Protein (g %)	2.7	15
Globulin (g %)	1.2	15
Albumin (g %)	1.5	15

TABLE 5. VALUES FOR CARDIOVASCULAR AND RESPIRATORY FUNCTION OF THE RABBIT

Parameter	Typical value	Reference(s)
Respiratory rate (breaths/min)	32–60	7
Heart rate (beats/min)	200–300	7
Tidal volume (ml/kg)	4–6	7
pO_2 (mmHg)	85–102	16
pCO_2 (torr)	20–46	7,16
HCO_3^- (mmol/l)	12–24	7
Arterial oxygen (% volume)	12.6–15.8	16
Oxyhemoglobin (%)	91–93	16
Arterial systolic pressure (mmHg)	90–130	7,17
Arterial diastolic pressure (mmHg)	80–90	7,17
Arterial blood pH	7.2–7.5	7,16

Cardiovascular and Respiratory Function

Cardiovascular and respiratory function is often altered with anesthesia or disease. Normal values of cardiovascular and respiratory function are presented in Table 5.

Hematology

Approximate ranges for hematologic parameters are shown in Table 6. Some variation of values can be expected to occur between individual rabbits, breeds, and laboratories. It is imperative that individual laboratories establish normal values for their facility. Noteworthy hematologic features include:

- The rabbit neutrophil contains eosinophilic granules and is commonly referred to as the **heterophil** or **pseudoeosinophil.** It is often mistaken for the similarly-appearing

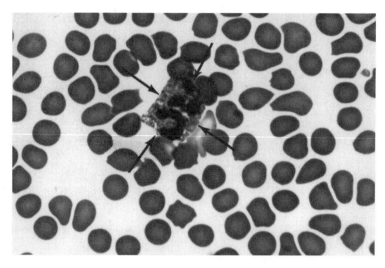

FIG. 2.   Heterophils (at the center of arrows). Note the granules in the cytoplasm. The two nuclei indicate two closely adjoined heterophils. Magnified ×1000.

eosinophil, although the heterophil is distinguished by smaller, lighter granules (Figure 2).

- Hyposegmented neutrophils can occasionally be microscopically observed in rabbit blood samples. This condition is known as the **Pelger-Huet** anomaly and is inherited as a partial dominant trait in rabbits.

Reproduction

Although high quality rabbits are readily available from commercial vendors, they can be bred with relative ease. Detailed information on the reproductive biology of rabbits can be found elsewhere.[19] Important aspects of rabbit reproduction include:

- Puberty usually occurs between 5 to 7 months of age in New Zealand White rabbits. For does, the breeding life lasts an average of 1 to 3 years, although some remain productive for up to 5 to 6 years. Bucks typically remain reproductively useful for 5 to 6 years.

- Females do not have a distinct estrous cycle, but rather demonstrate a rhythm with respect to receptivity, punctuated by periods of anestrus. Receptivity is often signaled

TABLE 6. HEMATOLOGIC VALUES OF THE RABBIT

Parameter	Typical value	Reference(s)
Packed cell volume (%)	34–43	7,10,12,18
Red blood cells (10^6/µl)	5.3–6.8	7,10,12,18
White blood cells (10^3/µl)	5.1–9.7	7,10,12
Hemoglobin (g/dl)	9.8–14.0	7,10,12,18
Neutrophils (%)	25–46	7,12
Lymphocytes (%)	39–68	7
Eosinophils (%)	0.1–2.0	7,12
Basophils (%)	2.0–5.0	7,12
Monocytes (%)	1.0–9.0	7,12
Platelets (10^3/µl)	158–650	7,12
Reticulocytes (%)	1.9–3.8	7
MCV (fl)	60–69	7,18
MCHC (%)	31–35	7,12
MCH (pg)	20–23	7,18
Blood volume (ml/kg)	55–65	7
Plasma volume (ml/kg)	28–50	7

by the vulva becoming swollen, moist, and dark pink or red.

- Bucks may be mated to more than one doe. Commonly, a single buck is sufficient to service 10 to 15 does.

- The doe is brought to the buck's cage for breeding, since the female rabbit can be extremely territorial and may attack the male.

- Ovulation is induced and occurs approximately 10 to 13 hours after copulation. Ovulation can also be induced by administration of luteinizing hormone,[20] human chorionic gonadotropin,[21] or gonadotropic releasing hormone.[22]

- Pregnancy can often be confirmed as early as day 14 of gestation by palpation of fetuses within the uterus, and radiographically by day 11.

- A nesting box with bedding material such as straw or clean, shredded paper should be provided to the doe several days prior to the expected kindling date. The doe will usually line the box with her own hair. Boxes of approximately 20 in. × 11 in. × 10 in. size are useful for

TABLE 7. REPRODUCTIVE SYSTEM VALUES OF THE RABBIT

Parameter	Typical value	Reference(s)
Gestation length (days)	31–32	7
Litter size (no. kits)	7–9	7
Birth weight (g)	30–100	7

this purpose. Care should be taken not to place the nesting box in the corner of the cage or pen where the doe typically urinates.

- A number of breeding strategies have been tried with success, including natural breeding, artificial insemination, and embryo transfer.[19,23,24]

- Average gestation length and litter and kit size are summarized in Table 7. Values shown are typical of the New Zealand White rabbit and can vary between breeders.

- Kindling (birth) normally occurs during the morning and takes approximately 30 to 60 minutes.[25]

- Both anterior and breech presentations are normal.[25]

- The doe normally nurses the kits once daily for only several minutes. Orphaned kits can be fed kitten milk replacer.[25]

- Nursing may last 5 to 10 weeks. Kits may begin consuming solid food by 3 weeks of age, with weaning generally occurring by 5 to 8 weeks of age.

- The ears of kits open at 8 days of age and the eyes at 9 to 11 days.

husbandry

Good husbandry is the cornerstone of a successful laboratory animal care and use program. An effective husbandry program should address the housing, social, and nutritional needs of the rabbit[26] as well as issues related to sanitation and record keeping.

housing

Housing of domestic rabbits should take into consideration both the **macroenvironment** (the space involving the entire room) and the **microenvironment** (the area immediately surrounding the animal).

Macroenvironment
The macroenvironment consists of the following elements:

1. **Location**. An optimal location for a room housing rabbits will be located at a site:

   - Where minimal noise is present from sources such as dogs or cagewash operations.

   - In close proximity to investigators and personnel responsible for care of the rabbits.

   - Which facilitates movement of caging equipment to and from the cage washing area.

2. **Equipment.** Equipment within the rabbit room should be:

   - Essential for daily operations related to the care and research use of the animals.

   - Kept out of the reach of the rabbits and stored in a manner which facilitates sanitation.

3. **Construction materials.** The following points should be considered regarding materials used for construction of the rabbit room:

   - All surfaces of floors and walls must be capable of withstanding disinfection procedures.

   - The floors should allow movement of cages with minimal effort.

   - The doors should be made of a material that can withstand cleaning and provide security for the animals housed within the room.

4. **Environmental control.** It is important that environmental variables remain consistent throughout the duration of a study.

5. **Design.** Rooms housing rabbits should be configured in such a way that:

   - Sufficient space is available so that cages are easily maneuvered and operations such as cage changing are easily performed.

   - The slope of the floor is toward the drain, yet not so steep as to interfere with stability of the rabbit cages.

Microenvironment

The caging system for the rabbits should provide safe and secure quarters for the animals and facilitate feeding and waste removal. Typically, several cages are secured on a cage rack equipped with casters (Figure 3). Appropriate caging should take into consideration the following:

Fig. 3. Rabbit cage rack with six cages.

Table 8. Cage Sizes for Rabbits

Body weight lbs. (kg)	Floor area/rabbit ft^2 (m^2)	Interior cage height in. (cm)
<4.4 (<2)	1.5 (0.14)	14 (35.56)
4.4–8.8 (2–4)	3.0 (0.28)	14 (35.56)
8.8–11.9 (4–5.4)	4.0 (0.37)	14 (35,56)
>11.9 (>5.4)	5.0 (0.46)	14 (35.56)

Summarized from the *Guide for the Care and Use of Laboratory Animals*[27] and the Regulations of the Animal Welfare Act.[28]

1. **Size.** Cages should be large enough to allow the rabbit to freely move about and stretch out when lying down. In addition, attention should be paid to regulatory specifications for cage size. For example, rabbit cage size standards as described in the *Guide for the Care and Use of Laboratory Animals*[27] and the Regulations of the Animal Welfare Act[28] are shown in Table 8. Standards from both of these sources are the same.

FIG. 4.   Cage door with a sliding latch.

2. **Cage Materials.**   Cages should be constructed of materials which are resistant to corrosion, strong enough to support and contain the animal, and facilitate disinfection. Stainless steel fulfills all these needs and is the most widely used material for rabbit cages. Heavy duty plastics have also been used successfully for rabbit caging.

3. **Design.**   Important cage design features include:

   • **Doors** which should open easily for personnel, yet be secure enough to prevent escape of the rabbit. In this regard, a sliding latch on the outside of the door is effective (Figure 4).

   • **Walls and tops** of cages, which vary between solid and mesh styles; however the openings must be sufficient to allow proper ventilation and observation.

   • **Wall and floor junctions** which should be rounded and lack crevices so as to minimize accumulation of waste materials.

   • **Floors** which may be slatted, expanded metal mesh, or solid bottom with perforated holes for removal of feces and urine. In the case of wire mesh, it is recommended that the openings be 1 inch by 1/2 in.[29]

- **Watering devices**, which include metal or ceramic bowls, bottles attached to the cage exterior, and automatic watering systems. Rabbits often overturn bowls or contaminate the water contained therein with urine or food material. Water bottles are effective, but care must be taken to assure that the bottle is firmly seated in its holder and that the sipper tube protrudes far enough into the cage. Automatic watering systems are less labor intensive, but adequate water flow from the sippers needs to be verified daily.

- **Feeding devices**, including bowls and feed hoppers attached to the cage exterior, are used. To minimize contamination by urine or feces, feed should not be offered on the bottom of the cage.

environmental conditions

It is important that the environment be maintained under stable conditions which promote the health of the rabbit. In this regard, important variables to be controlled include:

1. **Temperature.** Rabbits tolerate a fairly wide range of ambient temperatures, particularly cool temperatures. The recommended ambient temperature range for rabbits is 60.8 to 69.8°F.[27]

2. **Illumination.** Although the optimal lighting conditions for rabbits are unknown, common practice employs a 12- or 14-h light to a 12- or 10-h dark cycle. Light should be sufficient to allow daily observation of rabbits within their cages. Standard illumination intensities of 75 to 100 foot candles have been associated with retinal damage in albino rodents,[30,31] thus housing at lower light intensity may benefit the New Zealand White rabbit. Nonalbino breeds of rabbits are presumably less likely to develop similar abnormalities under standard light intensities.

3. **Ventilation.** Ventilation should be adequate to minimize room ammonia concentration since levels >50 ppm can increase susceptibility of rabbits to respiratory disease caused by *Pasteurella multocida*.[32] For this reason, recirculation of room air is not advisable. Ventilation rates of 10 to 15 room changes per hour are commonly employed for rabbit rooms.[27]

4. **Humidity.** Recommended values for relative humidity are 40 to 60%.[27]

5. **Noise.** Rabbits may be startled by sudden, sharp noise. Physiologic changes have been noted in rabbits exposed to elevated noise.[33] Personnel should exercise care to minimize noise in the presence of rabbits.

environmental enrichment

Typical caging situations provide limited opportunity for rabbits to interact socially or with their environment. In this regard, efforts to modify the environment to encourage increased interaction may include:

1. **Group housing.** Group housing allows rabbits the opportunity and space to more readily express their behavioral repertoire.[34] For example, group-housed female rabbits demonstrate behaviors such as mutual grooming and small-group socialization not possible in single-housed rabbits.[35] Group-penned female rabbits allowed to choose between single or paired housing prefer being in the same cage with other rabbits.[36] Group housing of male rabbits should be conducted with care to avoid fighting. Group-penned female and castrated male rabbits were more active, but agonistic behavior such as biting was noted even after dominance hierarchies were established.[37] Rabbits demonstrating aggression toward others should be separated from the group.

2. **Holding and petting by personnel.**

3. **Provision of novel objects**, such as empty, cleaned soft drink cans, balls, and wood blocks. Rabbits generally

interact with such objects by chewing.[36] All objects used in this manner should be free of sharp edges and be subject to minimal splintering. In addition, objects should be easily sanitizable or be replaced when soiled.

4. **Provision of novel food items**, such as broccoli, apples, and alfalfa. These items may supplement the normal feed ration; however, overfeeding of such treats may result in obesity.

nutrition

Rabbits are herbivorous animals and specific nutritional requirements depend upon age, health status, and reproductive state.[38,39] Information on rabbit nutrition is presented in greater detail elsewhere.[39] A number of high quality, nutritionally adequate commercial diets are readily available. The following points should be considered regarding nutrition of rabbits:

1. **Amount of feed.**   The recommended amount to be fed varies with the individual diet and rabbit. Young, growing rabbits may be fed *ad libitum*, whereas mature rabbits may need to be limit fed, especially if they are singly housed with limited opportunity for exercise. Experience has shown that **100–120 grams** of a typical commercial pelleted diet (Figure 5) is sufficient to maintain the health and weight of an average adult New Zealand White rabbit.

2. **Presentation of feed.**   Feed is normally offered to rabbits in a bowl or a J-feeder attached to the cage (Figure 6). It is important to offer the feed in such a way that contamination by urine or feces does not occur. For this reason, J-feeders are preferred over bowls placed on the cage floor. The feeder should be regularly cleaned to remove feed dust, which may decrease palatability of the food. In addition, feed may become caked and moldy at the bottom or bend of the "J" if the feeder is not thoroughly cleaned on a regular basis.

3. **Fiber.**   Although fiber digestibility in the rabbit is low,[40,41] fiber is an important component of the rabbit's diet and appears to play a role in prevention of problems

FIG. 5.   Pelleted feed.

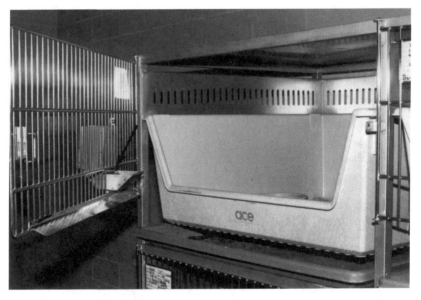

FIG. 6.   J-feeder (indicated by arrow) attached to cage door.

associated with the digestive tract. Recommendations for fiber range from 10% to 20% of the diet, although many diets contain fiber levels closer to the latter. Additional fiber can be provided in the form of alfalfa, apples, broccoli, or other palatable plant material.

FIG. 7. Rigid, calcified aorta (indicated by arrows) from a rabbit with hypervitaminosis D.

4. **Calcium.** In contrast to most other animals, the serum calcium level is not homeostatically regulated, but rises in direct proportion to increases in dietary calcium intake.[42] In addition, vitamin D plays a role in calcium absorption. Vitamin D toxicity has been reported in rabbits and is characterized by extensive calcification of soft tissues[43,44] (Figure 7). Urinary excretion of calcium is a major route of elimination in rabbits, whereas most other mammals excrete calcium via the bile.

5. **Vitamin A.** Both vitamin A deficiency and toxicity have been associated with reproductive problems in rabbits.[38,45,46] Disorders are characterized by fetal resorptions, abortions, stillbirths, and hydrocephalus. The precise vitamin A requirement for rabbits has not been determined, although 10,000 IU per kilogram of diet is adequate.[38]

6. **Vitamin E.** Vitamin E deficiency in rabbits has been associated with muscular dystrophy, abortions, stillbirths, and neonatal deaths in rabbits.[44,47,48] The precise vitamin E requirement of rabbits is not known.

7. **Storage of feed.** In general, commercial pelleted rabbit feed will retain acceptable nutritive levels for approximately 180 days after milling. For this reason, most reputable feed suppliers indicate the date of milling on the outside of feed packages. If the date is encoded, one may need to contact the feed supplier for code interpretation. Feed should be used no later than 180 days after the milling date.

Feed should be stored in a room which is neither excessively hot nor humid. Storage should be separate from storage of potentially hazardous substances such as insecticides or chemicals used for cleaning and disinfection. In addition, feed should be stored off the floor and away from the wall in order to facilitate sanitation (Figure 8). In this regard, plastic pallets are useful. Open bags of feed should be stored in containers with tightly fitted lids, to minimize vermin contamination. Storage procedures should facilitate display of specific information regarding the date of milling or the expiration date of the feed (Figure 9). Spilled feed should be promptly cleaned up and discarded.

8. **Coprophagy.** Rabbits normally produce and consume, at night, soft feces produced by fermentative processes in the cecum. This material is an important source of protein and B vitamins for the rabbit.

9. **Water.** Rabbits should be provided a constant source of fresh, clean, potable water. Water intake averages 50 to 100 ml/kg of body weight per day for rabbits, although growing, pregnant, and lactating rabbits require additional water. To some degree, water consumption is also influenced by the amount and type of feed, with increased water intake being associated with high fiber diets.[38] Water can be provided by means of water bottles attached to cages, automatic watering systems, or bowls placed in the cage.

sanitation

Effective sanitation is a crucial part of a sound husbandry program. The chief objective of sanitation is to maintain the cage

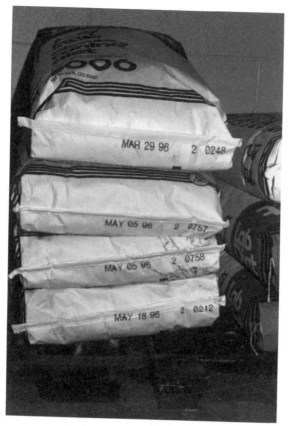

FIG. 8.   Storage of feed on plastic pallets. Note that bags of feed are stored so that the feed with the most recent milling date is located near the bottom to facilitate use of feed prior to expiration.

and room free of debris and waste that could provide a substrate for microorganisms and harborage for vermin. To achieve this objective, issues such as frequency and methods of sanitation must be considered.

Frequency

Frequency of cage and room sanitation may depend upon factors including the number of animals per room, efficiency of the ventilation system, and individual animal behavior. In general, experience has shown that effective sanitation may be achieved using the following schedule:

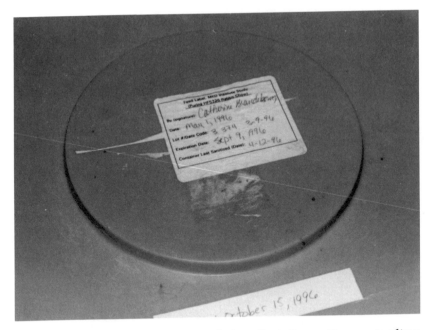

FIG. 9. Lid of feed barrel labeled with information regarding feed milling and expiration and date of last sanitization of the barrel.

1. **Daily** removal of feces and urine from the litter tray and sweeping or damp-mopping of animal room floors.

2. **Weekly** transfer of rabbits to clean, disinfected cages is optimal, although some facilities opt to change cages biweekly or monthly. Litter trays should be disinfected weekly, and litter from solid-bottom cages should be replaced weekly. These procedures should be performed more frequently if needed to provide adequate sanitation.

3. **Monthly** replacement of cage racks with clean racks.

4. **The animal room** should be cleaned at the conclusion of the study or whenever the room can be emptied of animals.

Methods

The following methods can be followed for disinfection of **caging equipment:**

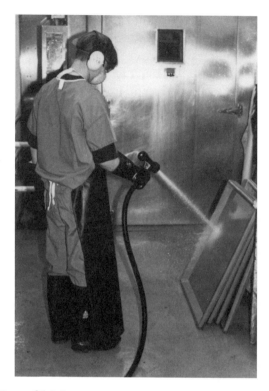

Fig. 10.   Use of high pressure water to remove waste material from a rabbit cage pan. Note that the operator is wearing appropriate protective equipment.

1. **Removal of all grossly visible debris** such as hair, feces, and urine scale (Figure 10). Brushing and rinsing with water are often sufficient to accomplish this step, although a detergent may be added to loosen or dissolve debris. Removal of urine scale may require washing or soaking equipment in a preparation containing an acid such as phosphoric acid or citric acid. Individuals performing these procedures should wear protective equipment to minimize the risk of chemical or thermal injury to the skin, eyes, and nasal passages. In addition, hearing protection is advisable, due to the intensity of noise which may be generated by cage-washing operations.

2. **Cleaning and disinfection** is generally performed by application of a chemical detergent and disinfectant followed by a water rinse. A variety of disinfectant chemicals

are available from a number of vendors. Any application of chemical to cages or equipment should be followed by thorough rinsing with water to minimize exposure of the animals to potentially harmful chemicals. Alternatively, hot water may be used to disinfect equipment. Typically, water at temperatures near 180°F is used for this step, although procedures using water at lower temperatures with longer sanitization times have been shown to be similarly effective.[49] Although sanitization may be effectively performed manually,[50] use of automatic cage washers are an efficient means to perform this function.

Note: Personnel should wear protective equipment such as protective gloves, aprons, and safety goggles when handling chemicals used for sanitation.

The following method can be used for disinfection of **animal rooms:**

1. **Removal of all gross debris** by rinsing with water and brushing. A detergent can be applied to help dissolve debris, followed by a water rinse.

2. **Application of a chemical disinfectant** followed by thorough rinsing with water.

Note: Animals should be removed from rooms or protected in such a way that they do not become wet or are exposed to chemicals during cleaning.

Quality Control

It is important that the overall efficacy of disinfection procedures be assessed periodically. Several methods may be employed in this regard:

1. **Temperature tape** monitoring utilizes tape applied to equipment undergoing thermal disinfection. The tape is designed to undergo a color change when exposed to temperatures consistent with disinfection (Figure 11). Monitoring cage washer temperature with temperature

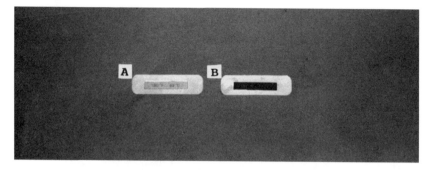

FIG. 11. Temperature tape before (A) and after (B) exposure to 180°F water.

tape is often performed each day the cage washer is used. Used tape samples can be maintained in a log as a record of acceptable cage wash temperature.

2. **Microbiologic monitoring** involves bacterial culture of equipment and room surfaces. Surfaces which might be tested include animal room surfaces, cages, cage racks, water bottles, sipper tubes, feeders, and any other equipment that undergoes disinfection. Swabs of such surfaces may be cultured or rodac plates can be gently touched against cage and room surfaces (Figure 12) and incubated. Moderate or heavy bacterial growth indicates weakness in the disinfection process and necessitates evaluation of disinfection procedures and equipment.

transportation

On occasion, live rabbits must be transported between facilities or institutions. Important aspects of transportation include the shipping container, provision of food and water during transit, observation of rabbits during transit, and environmental factors. In the U.S., specific requirements for transportation of rabbits are described in the Regulations of the Animal Welfare Act.[28]

1. **Shipping container.** The shipping container should be constructed of material durable enough to securely contain the rabbit and withstand damage that could result

Fig. 12. Application of a rodac plate to an interior wall of a cage.

in injury to it. Sufficient space should be provided to allow the rabbit to make normal postural adjustments. Containers should be designed to allow adequate ventilation of the container, yet protect the rabbit from drafts.

2. **Food and water.** The need to provide food and water during transportation depends upon the length of time and conditions of shipping. The Regulations of the Animal Welfare Act state that food and water need to be provided in instances when transit will last longer than 6 hours.[28] On the other hand, if relatively warm temperatures are anticipated during shipping, it may be beneficial to provide a water source for rabbits in transit for less than 6 hours.

3. **Observation.** Frequent observation of rabbits during transportation is recommended. The Regulations of the Animal Welfare Act require observation every 4 hours.[28] Adequacy of ventilation, ambient temperature, and general condition of the rabbit should all be assured during observation. Any factors resulting in stress to the animal

should be corrected as soon as possible. Animals developing health problems should receive prompt veterinary care.

4. **Environment.** Maintaining the environment that surrounds the rabbit during shipment is of great importance. In this regard, the ventilation system should provide fresh air with minimal drafts. Extremes of temperature should be avoided. In addition, provisions should be made for minimizing buildup of wastes in the shipping container. Addition of absorbent bedding will aid in waste management during shipping.

record keeping

Accurate records are of vital importance to a good husbandry program. Records should be maintained in such a manner that they are easily accessible to and understood by those who need them and easily completed by those annotating them. In addition, records should be maintained in such a way to protect them from theft, animals, or inadvertent damage from factors such as as moisture. It is important that all personnel clearly understand what specific information is to be recorded and the need for such records.

Identification of individual rabbits facilitates accurate record keeping procedures. Common methods used to identify individual rabbits include cage cards, plastic ear tags (Figure 13), tattoos (Figure 14), and implantable electronic microchip devices. Many vendors of research rabbits supply animals individually identified by one of these means. In the case of some vendors, breeding records based on rabbit identification can be used to determine the pedigree of individual animals if needed.

Following are some of the basic records that might be maintained by an animal research facility:

1. **Health records.** Daily records of animal health facilitate good animal care. Whether one keeps health records for individual animals or by room of animals is a matter of individual preference and requirements of the research. Routine entries should note the rabbit's state

FIG. 13. Plastic ear tag affixed to the right ear of a rabbit.

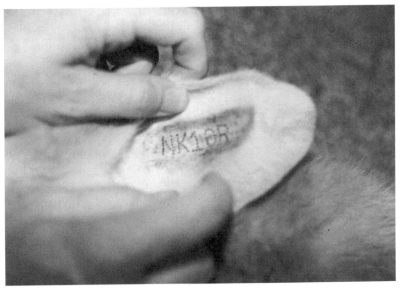

FIG. 14. Identification tattoo on the inner surface of the ear of a rabbit.

of activity and alertness, appetite, and specific abnor-malities. Additionally, experimental manipulations and administration of medications should be recorded. Infor-mation regarding abnormal animals should be promptly communicated to the attending veterinarian.

2. **Census.** Information concerning the number of animals in individual rooms and the overall facility is helpful in planning the daily work load. Census information main-tained on a room card on the outside door and logged into a permanent record is often useful.

3. **Work records.** Records of routine husbandry tasks car-ried out in animal rooms should be maintained. Basic relevant information includes food and water provision and intake, changing of cages and cage rack, tempera-ture, date, and time of day when the animals are checked, and initials of the individual logging the infor-mation.

notes

management

regulatory agencies and compliance

Specific regulatory agencies and requirements may vary with locale; however in the U.S. the following are the primary organizations with regulatory oversight or accreditation responsibilities for programs of research, teaching, or testing involving rabbits:

U.S. Department of Agriculture

- Oversight responsibility is described in the **Animal Welfare Act** (P.L. 91-579, 94-279, 99-198).[51]

- Specific regulatory requirements are described in the **Regulations of the Animal Welfare Act**.[28]

- Registration with USDA and adherence to USDA regulations are required by all institutions, except elementary or secondary schools, using rabbits in teaching, testing, or research in the U.S.

National Institutes of Health, Public Health Service (PHS)

- Oversight responsibility is described in the **Health Research Extension Act of 1985** (P.L. 99-158).[52]

- Policy is described in the **Public Health Service Policy on Humane Care and Use of Laboratory Animals**.[53]

- Adherence to the PHS Policy is required of those institutions conducting research using funds from PHS.

- Principles for implementation of PHS policy are those described in the *Guide for the Care and Use of Laboratory Animals*.[27]

U.S. Food and Drug Administration (FDA) and the Environmental Protection Agency (EPA).

- Policies are described in the **Good Laboratory Practices for Nonclinical Laboratory Studies** (CFR 21 (Food and Drugs), Part 58, Subparts A–K; CFR Title 40 (Protection of Environment), Part 160, Subparts A–J; CFR Title 40 (Protection of Environment), Part 792, Subparts A–L).

- In general, standard operating procedures must be outlined and rigorously followed and supported with detailed records.

- Adherence is required when using rabbits in studies used to request research or marketing permits as part of the approval process for drugs or medical devices intended for human use.

Association for Assessment and Accreditation of Laboratory Animal Care International, Inc. (AAALAC International)

- AAALAC International is a nonprofit organization designed to provide peer review-based accreditation to animal research facilities.

- Basis for accreditation is adherence to principles described in the **Guide for the Care and Use of Laboratory Animals**.[27]

- Accreditation is voluntary.

In addition to the above regulatory bodies, state and local regulations may exist.

institutional animal care and use committee (IACUC)

The basic unit of an effective animal care and use program is the Institutional Animal Care and Use Committee (IACUC). The USDA, Public Health Service (PHS), and AAALAC-International require an IACUC at any institution using rabbits in research, teaching, or testing. Important points regarding the composition of the IACUC include:

1. **Number of members.** USDA regulations require a minimum of three members, while the PHS policy requires a minimum of five members.

2. **Qualifications of members.** The IACUC should include the following:

   - A chairperson

   - A doctor of veterinary medicine who has training or experience in laboratory animal medicine or science, and responsibility for activities involving animals at the research facility.

   - An individual who is in no way affiliated with the institution other than as an IACUC member. At some institutions this role has been fulfilled by clergypersons, lawyers, or local humane society or animal shelter officials.

In addition, PHS policy requires the following members:

   - A practicing scientist with experience in animal research.

- One member whose primary concerns are in a non-scientific area. This individual may be an employee of the institution served by the IACUC.

It is acceptable for a single individual to fulfill more than one of the above categories.

Responsibilities of the IACUC

The written regulations should be consulted for an in-depth description of IACUC responsibilities. In general, the IACUC is charged with the following:

- To review proposed protocols for activities involving use of animals in research, teaching, and testing. Protocols must be approved by the IACUC before animal use may begin.

- To inspect and assure that the animal research facilities and equipment meet an acceptable standard.

- To assure that personnel are adequately trained and qualified to conduct research using animals.

- To assure that animals are properly handled and cared for.

- To assure that the investigator has considered alternatives to potentially painful or stressful procedures and has determined that the research is nonduplicative.

- To assure that sedatives, analgesics, and anesthetics are used when appropriate.

- To assure that proper surgical preparation and technique are utilized.

- To assure that animals are euthanized appropriately.

occupational health and zoonotic diseases

Domestic rabbits purchased from reputable vendors pose virtually no risk of infectious zoonotic disease, unless experimentally infected with zoonotic pathogens. The need for and aspects of comprehensive programs for occupational safety and health for individuals working with laboratory animals have been described.[54] In general, personnel should wear a clean lab coat or coveralls when working with research animals. Occupational health programs for personnel handling rabbits should be developed with consideration for the following potential health issues:

1. **Puncture, bite and scratch wounds.** Rabbits will infrequently bite; however, scratch wounds, while often inflicted unintentionally, are considerably more common. Puncture wounds may result from handling equipment with sharp edges or points. For this reason, personnel should be current with respect to tetanus immunization.

2. **Ringworm.** Ringworm results from infection of cutaneous tissues with fungi called dermatophytes. Any rabbit demonstrating characteristic ringworm lesions, as described in the veterinary care section (Chapter 4) of this handbook, should be considered as a possible source of human infection.

3. **Allergy.** Allergies to rabbit dander are not uncommon in personnel exposed to rabbits. Personnel may experience respiratory symptoms such as sneezing and rhinitis or skin symptoms such as redness, swelling, and pruritis following exposure. As with many other allergies, extreme sensitivity to rabbit dander can result in anaphylaxis and thus represents a serious occupational hazard for some individuals. It is advisable for sensitive personnel to wear a face mask or fitted respirator, gloves, and a clean, launderable lab coat or coveralls (Figure 15). Ideally, sensitive personnel should be reassigned to job tasks which eliminate the possibility of exposure to allergens. The advice of an occupational health specialist should be sought and followed if reassignment away from rabbit areas is not possible.[55] In addition, it is advisable for such individuals to undergo periodic respiratory function testing.

4. **Experimental biohazards.** Some studies may involve purposeful infection of rabbits with known human pathogens. In such cases, it is recommended that standard operating procedures for safe handling of biohazardous

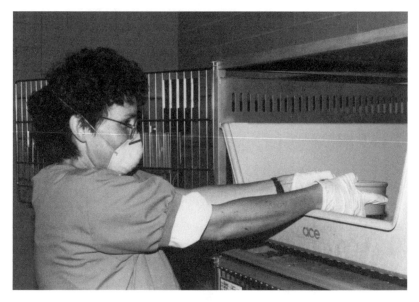

FIG. 15. Animal caretaker wearing disposable gloves, mask, and launderable coveralls to decrease exposure to allergens.

materials and infected animals be established and followed. Guidelines for use of biohazardous agents are discussed in detail elsewhere.[56]

veterinary care

basic veterinary supplies

The following basic supplies are useful for the clinical care of rabbits:

1. A stethoscope

2. Disposable syringes, ranging in size from 1 ml to 12 ml

3. Disposable hypodermic needles, ranging in size from 21 to 26 gauge (diameter) and 5/8 to 1 1/2 in. (length)

4. Blood collection tubes with no additive (for serum) or added EDTA (for whole blood)

5. Gauze sponges

6. Small animal rectal thermometer

7. Lubricating jelly

8. Disinfectant such as povidone-iodine solution

9. Sterile fluids such as lactated Ringer's solution or 0.9% sodium chloride

10. Nail clippers

11. Bacterial culture swabs in transport media

12. 8 French infant feeding tubes for oral gavage

Additional supplies should supplement those listed above, depending upon the needs of the facility.

physical examination of the rabbit

A physical examination should be performed on rabbits upon arrival at the facility and on rabbits exhibiting any abnormalities. Findings should be recorded in the medical records for the animal. Physical examination of the rabbit is performed in the following manner:

- General assessment of behavior of the animal within the cage and during removal from the cage. Findings such as lethargy or aggressiveness should be noted.

- The feces and urine in the cage pan should be inspected, and abnormal consistency, color, or odor noted.

- The feed bowl or hopper should be checked to evaluate the rabbit's appetite.

- The coat should be examined for hair loss, open or closed skin lesions, or abnormal masses.

- The bottom of the feet should be examined for open or closed lesions. Overgrown toenails should be clipped.

- The eyes should be examined for discharge or abnormal reddening of the conjunctiva.

- The nose should be examined for discharge.

- The lips and mouth should be examined for lesions and overgrown incisors.

- The ears should be examined for accumulation of dry, brown, crusty material, which is suggestive of ear mite infestation.

- A stethoscope should be used to listen for abnormal respiratory sounds, which may indicate pneumonia.

- The abdomen should be palpated for abnormal masses within the abdominal cavity. This is performed by standing directly behind the rabbit and firmly pressing the fingers of both hands into the cranial part of the abdomen

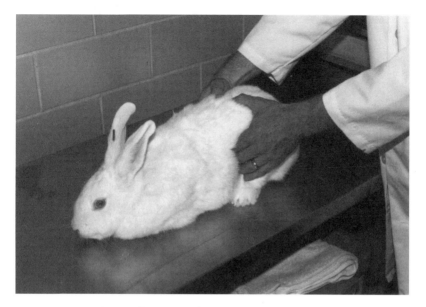

FIG. 16.   Palpation of the abdomen of a rabbit.

and slowly drawing the fingers back caudally, being sure to palpate both ventral and caudal aspects of the abdomen (Figure 16).

- The perineal region is examined for fecal or urine staining, vulvar discharge, and for open or closed lesions.

- The body temperature may be measured rectally by inserting for 2–3 minutes a small animal glass rectal thermometer which has had a small amount of lubricating jelly applied to the bulb. Alternatively, the body temperature may be measured from the ear by use of an infrared tympanic thermometer, although the accuracy of this type of thermometer has not been validated for rabbits. The normal body temperature for rabbits, as measured by rectal thermometer is 38 to 40°C (100.4 to 104.0°F), with some variation related to ambient environmental conditions.

quarantine

Groups of rabbits arriving from the vendor should be isolated for 7 to 10 days. Although physiologic stabilization occurs within

Fig. 17. Rabbit with ocular discharge. Note the overgrowth of the incisors of this rabbit.

48 hours,[57] subclinical disease may require longer periods to demonstrate clinical manifestations.

common clinical problems

Rabbits can develop a variety of clinical problems, as described below. Many of these diseases can be avoided by using specific pathogen-free (SPF) rabbits from reputable vendors.

General Signs Suggestive of Illness

- Loss of appetite and weight loss
- Unkempt appearance, hair loss or thinning
- Lethargy and inactivity
- Nasal or ocular discharge (Figure 17)
- Diarrhea (Figure 18)

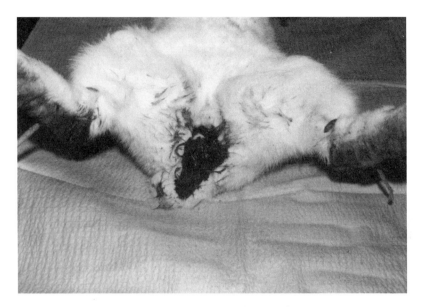

FIG. 18.   Rabbit with diarrhea.

A summary of clinical disease associated with specific etiologic agents is presented in Table 9, and additional information for some of these diseases and agents follows.

Control of infectious disease in rabbits involves strict sanitation and isolation or removal of sick animals. Ideally, rabbits should be housed in such a way that contact between SPF rabbits and rabbits harboring potential infectious pathogens is minimized. Often, this is best achieved by maintaining them using barrier facilities and procedures.[58] Such procedures generally involve maintenance of the room under positive air pressure with high efficiency particulate air (HEPA)-filtered air, restriction of access to only those personnel essential to the care and use of the rabbits, donning of disposable gloves and clean laboratory coats or coveralls which are worn only in the barrier, requirements that any rabbits leaving the barrier are not permitted to return to the barrier, thorough disinfection of supplies before entry into the barrier, and, in some cases, use of autoclavable feed.

Note: The risk of infectious disease can be minimized by using specific pathogen-free (SPF) rabbits exclusively.

TABLE 9. COMMON DISEASES OF RABBITS

Clinical signs	Etiololgic syndrome
Diarrhea, distended abdomen, anorexia	Coccidiosis
Diarrhea, anorexia, high death rate	Clostridial enterotoxemia
Nasal and/or ocular discharge, anorexia, labored breathing, abscessation, reduced reproductive ability, head tilt and circling	Pasteurellosis
Ulcers and crusts on nostrils, mouth, eyelids, vulva, and prepuce	Venereal spirochetosis
Bloody vulvar discharge, anorexia, weight loss, reproductive failure	Uterine adenocarcinoma
Patchy hair loss, thinning of hair	Hair pulling
Crusty, pruritic hairless lesions on face or ears	Dermatophytosis
Ulcerated lesions on bottom of feet, reluctance to move	Ulcerative pododermatitis
Accumulation of brown, crusty matter in ear, head shaking, ear scratching	Ectoparasitic otitis externa
Hind limb paresis/paralysis	Lumbar spinal fracture
Anorexia, weight loss, ptyalism	Dental malocclusion
Sudden anorexia, reduced fecal output	Trichobezoar
Enlargement of the eye	Buphthalmia

The diseases and treatment/prevention strategies discussed here represent the problems most likely to be seen in a typical laboratory rabbit. More complete reviews of rabbit diseases can be found elsewhere.[3,59,60] Treatment of rabbits for disease should be implemented under the direction of a qualified veterinarian, following appropriate diagnostic measures. A list of drugs commonly used in rabbits is provided in Table 10.

Coccidiosis

- Caused by several species of the protozoan parasite, *Eimeria*, including *E. stiedae* (hepatic isolate), *E. magna* (intestinal isolate), and *E. irresidua* (intestinal isolate).

- Infection may be subclinical or cause mild to severe diarrhea.

- If the liver is affected, the abdomen may have a pendulous appearance. Numerous yellow or white spots are often present on and in the liver (Figure 19).

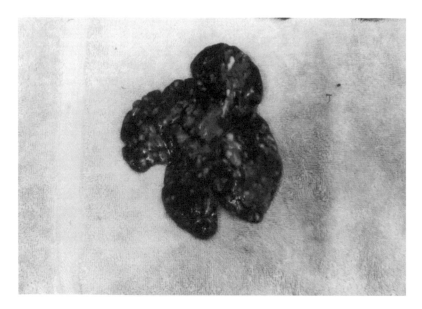

Fig. 19. Liver from a rabbit infected with *Eimeria stiedae.*

- Diagnosis is by examination of fecal flotation preparations for coccidial oocysts.

- Various sulfonamides will ameliorate, but not eliminate infection.[61]

- Control involves strict sanitation.

Clostridial Enterotoxemia

- Caused by the spore-forming, toxin-producing, anaerobic bacteria, *Clostridium spiroforme.*[62]

- Weanlings are more susceptible than adults.

- Causes moderate to severe diarrhea frequently resulting in death.

- Diagnosis is by histopathologic examination of the cecum and colon for inflammation, isolation of *C. spiroforme* under strict anaerobic culture conditions, microscopic examination of feces for coiled Gram-positive organisms, or by assay of cecal contents for *C. spiroforme* iota toxin.[63]

- Feeding of diets with high fiber (approximately 20% fiber) may serve to prevent the disease.

- Cholestyramine administered daily by gavage (2 g/20 ml of water) may help to prevent death of individuals in the event of an outbreak.[64]

- Rigorous disinfection can help to control the disease.

Pasteurella multocida Infections

- Gram-negative bacteria; the most common bacterial pathogen of rabbits.

- Infection may be subclinical or cause various combinations of clear to thick nasal discharge, clear to thick ocular discharge, anorexia, lethargy, and respiratory distress due to pneumonia.

- Infection may also result in abscesses in subcutaneous tissues and other sites (Figure 20), torticollis (head tilt) and circling related to infection of the inner ear (Figure 21), and septicemia.

- Diagnosis is by bacterial culture of the nasal cavity or other affected tissues.

- Antibiotic treatment may include penicillin,[65] gentamicin,[65] and chloramphenicol.[66] Some investigators have found enrofloxacin to be highly effective at elimination of *P. multocida* infection,[66,67] while others have found it to be ineffective.[68] Although *P. multocida*-free kits can be derived by treatment of infected does with enrofloxacin,[69] use of enrofloxacin in pregnant does may not be advisable, since the drug readily crosses the placenta[70] and is excreted in the milk[71] in rabbits. In other species, enrofloxacin has been associated with cartilage damage in the young.[72,73]

- Eradication of *P. multocida* from an entire colony can be attempted by various management and therapeutic procedures.[74]

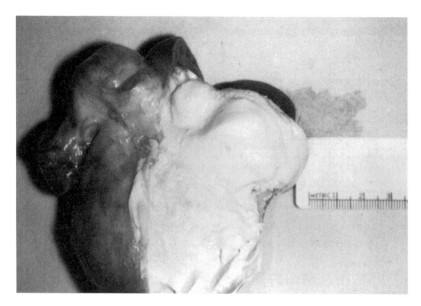

FIG. 20. Incised kidney with *P. multocida* abscess. Note the thick, white inflammatory exudate which is typical of such abscesses.

FIG. 21. Torticollis (head tilt) due to inner ear *P. multocida* infection.

Venereal Spirochetosis

- Caused by the spirochete bacteria, *Treponema cuniculi.*

- Transmitted by direct contact with an infected rabbit.

- Infection may be subclinical or may result in raised, crusted or ulcerated lesions on the genitalia, perineal region, nose, mouth, and eyelids.

- Diagnosis is by darkfield microscopic examination of lesion scrapings.

- Condition may resolve spontaneously within several weeks to months or it can be treated with a combination of benzathine penicillin and procaine penicillin.[75]

Uterine Adenocarcinoma

- Spontaneous, highly metastatic neoplasm common among does older than 2 years.

- Clinical signs include anorexia, weight loss, reduced reproductive performance, and bloody vulvar discharge. Labored respiration can result from metastasis to the lungs.

- Diagnosis is usually by palpation of abnormal masses within the abdominal cavity.

- Affected animals should be humanely euthanized.

Hair Pulling

- Presumably due to nest building behavior, boredom, or seasonal moulting.

- Environmental enrichment strategies may relieve hair pulling related to boredom.

Dermatophytosis (Ringworm)

- In rabbits, most commonly caused by the fungus, *Trichophyton mentagrophytes.* Less commonly, *Microsporum canis* has been implicated.[76]

- Infection may be subclinical or characterized by hair loss, reddening of skin, and crusts or scabs on the face, ears, and forelimbs. Rabbits may vigorously scratch at lesions.

- Diagnosis is made by clinical appearance, culture of hairs at the lesion margins on dermatophyte test media (DTM), or by microscopic examination of lesion skin scrapings mounted in 10% KOH for typical dermatophyte organisms.

- Transmissible to humans.

- Treatment includes application of antifungal creams to lesions or systemic treatment with griseofulvin.[77] Affected rabbits can also be successfully treated with 1% copper sulfate applied as a dip or with a dilution of a metastabilized chlorous acid/chlorine dioxide compound applied as either a dip or a spray.[78]

Ulcerative Pododermatitis

- Frequent in heavy rabbits housed on wire mesh floors.

- Characterized by open, ulcerated lesions on the bottom of the feet (Figure 22). Although any of the feet can be affected, the rear feet are more commonly involved, since rabbits bear most of their weight on the rear limbs and will occassionally stamp their rear feet when they feel threatened.

- Treatment includes cleaning the lesion and applying topical antibiotic, and providing rabbits with a plexiglass resting board or other type of resting mat. Alternatively, affected rabbits may benefit from housing in a solid-bottom cage with wood shavings provided as bedding material.

Ectoparasitic Otitis Externa

- Caused by infestation with *Psoroptes cuniculi*, the rabbit ear mite.

- Characterized by presence of dry, brown, crusty material adherent to the inner surface of the ear. Rabbits may scratch their ears or shake their heads due to intense pruritis.

Fig. 22.   Rabbit with pododermatitis.

- Diagnosis is confirmed by observation of *Psoroptes* organisms upon microscopic exam of swabs of the accumulated crusty material. Mites are large and may be seen with the naked eye.

- Treatment may include instillation of 1 to 2 ml of mineral oil containing an insecticide or acaricide into the ear canal. Alternatively, ivermectin is effective.[79-81]

Fracture of the Lumbar Spine

- Relative strength of the hind limb musculature to the weakness of the lumbar spine predisposes rabbits to fracture of the lumbar spine when rabbits jump or thrash during handling.

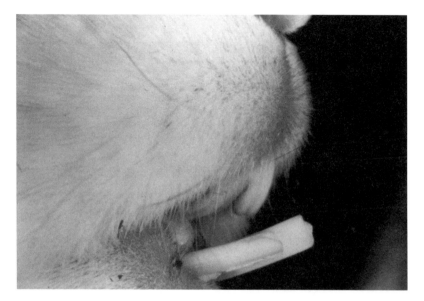

FIG. 23. Rabbit with overgrown incisors due to malocclusion.

- Characterized by partial or complete hind limb paresis or paralysis, loss of control of urinary bladder function, and diminished anal sphincter tone.

- Diagnosis can be confirmed by radiography.

- Affected animals generally should be euthanized. Extremely valuable animals can be maintained by providing intensive nursing care, including expression of the bladder several times daily, daily cleaning of the animal, provision of a resting cushion and frequent changing of the rabbit's position to prevent decubital ulcers, and assuring adequate ingestion of food and water.

Dental Malocclusion and Overgrown Teeth

- Continuously erupting teeth of the rabbit need to wear against one another. If occlusion is not proper, the teeth will overgrow (Figure 23).

- Due to inherited mandibular prognathia (most common cause) or to fractured or missing teeth.

- Characterized by weight loss, difficulty with mastication, and ptyalism.

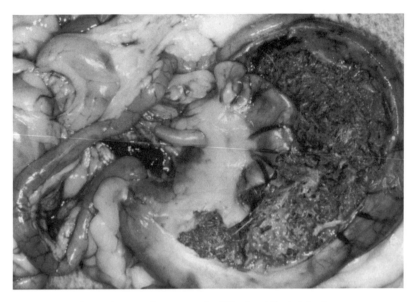

FIG. 24. Incised stomach filled with trichobezoar.

- Overgrown teeth should be periodically clipped or trimmed, preferably with a dental bur, although a toenail trimmers can be used. Care must be taken not to crack the tooth and to smooth any sharp edges.

Trichobezoar

- Fastidious grooming habits predipose rabbits to ingestion and subsequent accumulation of hair in the stomach or pylorus (Figures 24 and 25). Vomition of accumulated hair or other material by the rabbit is precluded by annular gastric musculature at the base of the esophagus.

- Most rabbits with gastric trichobezoars remain clinically normal,[82] however a few may experience anorexia, weight loss, and decreased fecal output if the trichobezoar obstructs the flow of ingesta to a significant degree. Death may result if the obstruction is prolonged.[83] Occassionally, gastric rupture with subsequent peritonitis may occur.[84]

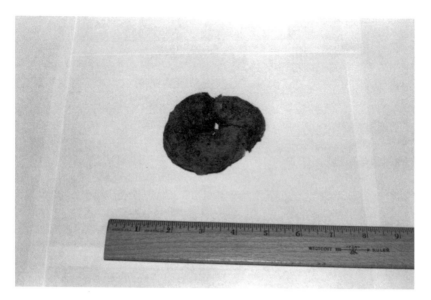

FIG. 25. Dried trichobezoar taken from the stomach of an anorexic rabbit.

- Diagnosis is confirmed by palpation of an abnormal, doughy mass in the cranial abdomen. Contrast radiography is not consistently conclusive.[85]

- Treatment may include administration of 10 to 15 ml of mineral oil by oral gavage (see Chapter 5 for oral gavage technique), although some authors suggest administering 5 to 10 ml of fresh pineapple juice for 3 to 4 days.[3]

- Surgical gastrotomy may be necessary in some cases; however rabbits which have been anorexic for extended periods often do not survive surgery due to their poor physiologic condition. In addition, gastrotomy may predispose the rabbit to subsequent reformation of a gastric trichobezoar.[82]

- High fiber diets may be useful in prevention of trichobezoar formation.[86]

Buphthalmia

- An inherited, autosomal recessive trait.

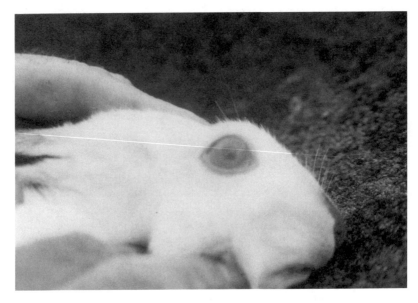

FIG. 26.   Cloudy cornea of a buphthalmic rabbit.

- Clinical signs may include an enlarged and bulging eye, increased size of the anterior chamber of the eye, bluish or cloudy appearance of the cornea, and conjunctivitis[87] (Figure 26).

- Affected animals may have increased intraocular pressure as early as 3 months of age.[88]

- The enlarged eye is subject to trauma and drying. Periodic application of ocular lubricants may lessen the likelihood of drying.

treatment of disease

Drug Dosages

Treatment of sick rabbits should be implemented under the direction of a qualified veterinarian, following appropriate diagnostic measures. A short list of drugs which can be used in rabbits is provided in Table 10. Dosages in Table 10 are expressed per unit of body weight as shown. Abbreviations for route of administration are PO (oral), IV (intravenous), IM (intramuscular),

TABLE 10. DRUGS DOSAGES FOR RABBITS

Drug	Dosage information	General application	Reference
Cephalexin	15–20 mg/kg PO/BID	Bacterial infections	89
Dexamethasone	0.5–2 mg/kg, IM	Shock, anti-inflammatory	59
Doxapram	2–5 mg/kg IV	Stimulation of respiratory activity	89
Enrofloxacin	5 mg/kg IM/BID	Bacterial infections	67
Gentamicin	5–8 mg/kg SC/SID	Bacterial infections	65,90
Griseofulvin	25 mg/100 g PO repeated in 10 days	Dermatophytosis	77
Ivermectin	200–400 µg/kg SC twice, 14-18 d apart	Ectoparasitism	78–81
Oxytocin	1–2 units IM or SC	Stimulation of uterine contractions	91
Prednisone	0.5–2 mg/kg, IM or SC	Anti-inflammatory	59
Procaine Penicillin G + Benzathine Penicillin G	84,000 IU/kg SC once weekly for three weeks	Venereal spirochetosis	75
Sulfamethazine	100 mg/kg PO	Coccidiosis	92

and SC (subcutaneous). Abbreviations for frequency of administration are SID (once daily), and BID (twice daily).

General Treatment of Diarrhea

Rabbits may quickly become dehydrated following diarrhea, and treatment should aim to replace lost fluid and electrolytes, regardless of the precise etiology. The severity of dehydration can be assessed by the "skin tent" test, in which a fold of skin on the dorsum of the neck is lifted and released. Ordinarily, the skin should return to its position within 1 to 2 seconds; however if the animal is dehydrated, the return to normal position will take longer.

In general, appropriate replacement fluids should be isotonic and contain an excess of bicarbonate (40 to 50 mEq/l), since diarrhea may precipitate acidosis. If the animal is also anorexic, dextrose may be added to provide caloric supplementation. It is important to assess hydration of the animal while it is receiving fluids, since over-admininstration of fluids can lead to serious complications. As a general guideline, fluids may be given in increments of 30 to 40 ml/kg of body weight until normal hydration is achieved. Fluids are most easily administered subcutaneously. Intravenous administration is also practical, and intraperitoneal administration is an alternative route.

Animals experiencing fluid and electrolyte loss often become hypothermic as they lapse into shock. For this reason, it is important to maintain the body temperature of animals with severe diarrhea. To this end, such rabbits can be provided abundant, clean bedding or towels. Supplemental heat can be provided by means of a heat lamp positioned a minimum of 10 to 12 inches from the animal or by means of a heat blanket. Close attention should be paid to animals on heat blankets, since malfunctioning blankets can cause severe thermal burns to animals. Heat blankets using a circulating warm water system provide a safe and effective source of supplemental heat.

If diagnostic efforts implicate a specific etiology as the cause of diarrhea, specific treatment directed against that etiology should be pursued, under the direction of a veterinarian.

General Treatment of Anorexia

Rabbits may experience anorexia for a number of reasons; however the specific cause is not always discernable. Although any specific cause should be appropriately eliminated following diagnostic identification, it is also appropriate to engage nonspecific measures to encourage rabbits to eat and thereby overcome anorexia. Some rabbits appear to have strong food preferences and can be coaxed into eating by offering them sliced apple, broccoli, alfalfa, carrots, carrot greens, or cabbage. Some rabbits prefer food covered with a few drops of molasses. Commercially available nutritional supplements (e.g., Nutrical®, Evsco Pharmaceuticals, Buena, NJ) can be administered orally. If the rabbit has ceased drinking, fluid supplementation may be necessary. Administration of 50 to 100 ml/kg/d of lactated Ringer's solution with dextrose can be given subcutaneously. Although anorexia will resolve quickly in some rabbits, others may consume a small amount of preferred foods for longer periods, even 1 to 2 weeks, before fully recovering. Animals which experience significant loss of body condition in the face of prolonged, unresolving anorexia should be humanely euthanized.

General Treatment of Open Skin Lesions

Rabbits may develop open skin lesions for a variety of reasons, including trauma, abscesses related to bacterial infection, and abscesses related to the use of Freund's complete adjuvant for polyclonal antibody production. In all cases, it is important to keep the lesion free of contaminating debris. In addition, it is advisable to clean all open lesions at least every other day with an antiseptic solution such as betadine.

Whether or not the inciting cause of the lesion was a bacterial pathogen, any open wound is susceptible to secondary bacterial infection. For this reason, application of a topical bacterial ointment should be considered. In addition, it is advisable to perform bacterial culture on lesions which demonstrate drainage suggestive of infection. Systemic antibiotic therapy should be based on the outcome of culture and antibiotic sensitivity testing.

It is unwise to attempt to suture or otherwise close draining lesions. To the contrary, drainage should be permitted as part of the normal healing process.

disease prevention through sanitation

Practicing proper sanitation is the best way to control many diseases of the rabbit. Rabbit cages should be routinely cleaned and disinfected as described in Chapter 2. Efforts should be made to prevent excessive accumulation of feces, urine, and dander, both in the cage and in the room. Instruments and equipment used on more than a single animal should be cleaned and disinfected between rabbits. In addition, use of disposable gloves will facilitate control of infectious disease. Personnel should wash their hands with an antiseptic soap after handling rabbits suspected of harboring infectious agents. Optimally, rabbits infected with pathogens should be isolated from noninfected animals.

anesthesia and analgesia

Procedures which may produce more than momentary pain should include the appropriate use of anesthetics and/or analgesics. **General anesthesia** is appropriate for highly invasive or otherwise painful procedures, while **local anesthesia** may be used when desensitization of a small, localized anatomic site is appropriate. Procedures which may produce pain lasting beyond the duration of anesthesia, such as those involving major surgery, should include the use of **analgesics** for postprocedural pain relief.

Principles of General Anesthesia

Procedures which produce more than momentary pain should involve the use of anesthetics. A number of commonly used anesthetic compounds and regimens are described below and briefly summarized for quick reference in Table 11. More detailed reviews of rabbit anesthesia can be found elsewhere.[93] Only the minimal amount of anesthetic needed to maintain an acceptable level of anesthesia should be used. Great variation may exist between individual rabbits in terms of the amount of anesthetic needed to induce a surgical depth of anesthesia and the duration of anesthetic effect. In general, the animal should be first given a low dose of the anesthetic and additional doses administered if needed. Anesthetics and analgesics are most appropriately used under the guidance of a qualified veterinarian.

Abbreviations for route of administration are PO (oral), IV (intravenous), IM (intramuscular), SC (subcutaneous), and IN (intranasal).

Characteristics of Commonly Used Injectable Anesthetics

Ketamine hydrochloride

- Dissociative anesthetic.

- Produces poor muscle relaxation, copious salivation, and lacrimation.

- May produce moderate decreases in cardiopulmonary function when used with xylazine.[101,102]

- Rabbits may self-mutilate if ketamine/xylazine is injected intramuscularly too close to major nerves such as the sciatic.[103]

- Yohimbine (0.2 mg/kg, IV) can be used to reverse ketamine/xylazine anesthesia.[104]

TABLE 11. COMMON ANESTHETIC DRUGS FOR RABBITS

Agent	Dosage/route	Approximate duration	Reference
Thiopental	15–30 mg/kg, IV	5–10 min	94
Pentobarbital	20–40 mg/kg, IV	30–45 min	95
Ketamine + xylazine	35 mg/kg + 5–10 mg/kg, IM	35–90 min	96,97
Ketamine + xylazine + acepromazine	35 mg/kg + 5 mg/kg + 0.75 mg/kg, IM	60–100 min	96
Ketamine + xylazine + butorphanol	35 mg/kg + 5 mg/kg + 0.1 mg/kg, IM	80–100 min	98
Tiletamine/zolazepam + xylazine	15 mg/kg + 5 mg/kg, IM	70 min	97
Propofol	1.5 mg/kg IV, then 0.2–0.6 mg/kg/min by IV infusion	Varies with time of infusion	99
Tiletamine/zolazepam	10 mg/kg, IN	45 min	100
Ketamine + midazolam	25 mg/kg + 1.0 mg/kg, IN	50 min	100
Ketamine + xylazine	10 mg/kg + 3 mg/kg, IN	35 min	100
Halothane	1.5–2.0% given by inhalation	Varies with length of exposure	93
Methoxyflurane	0.4–1.0% given by inhalation	Varies with length of exposure	93
Isoflurane	1.0–3.0% given by inhalation	Varies with length of exposure	93

Xylazine

- Alpha-2 adrenergic agonist.

- Produces good muscle relaxation.

- Produces depression of cardiac output and hypothermia. Death due to acidosis following ketamine/xylazine anesthesia can occur in rabbits.[105]

Tiletamine/zolazepam

- Tiletamine is similar to ketamine, while zolazepam is a diazepinone tranquilizer.

- May cause renal disease when used alone at anesthetic levels,[106] but lower dosages combined with xylazine are safe.[97] The nephrotoxicity of the drug has been attributed to the tiletamine component.[107]

- Lower dosages administered intranasally are effective.[100]

- Produces good muscle relaxation, but decreased cardiac output.

Pentobarbital and thiopental

- Barbituric acid derivatives.

- Thiopental is ultra-short acting and is useful for short procedures, including introduction of an endotracheal tube for administration of inhalation anesthesia. Longer-lasting anesthesia may be produced by means of a slow thiopental IV infusion drip.

- Pentobarbital produces longer-lasting anesthesia than thiopental and may be given as a bolus or a slow IV drip.

- Both produce good muscle relaxation, but significant respiratory depression, hypothermia, and hypotension.

- Both should be administered slowly, since some animals may succumb quickly to the adverse effects of the drug and may not need the total calculated dose.

- Always administer barbiturates **intravenously**, since IM or SC administration may result in irritation and possibly necrosis at the injection site.

Propofol

- Substituted phenol derivative.

- Induces short-duration anesthesia following IV administration;[108] good for quick procedures, including endotracheal intubation.

- Propofol should not be used for long-term anesthesia due to adverse effects on the cardiopulmonary system.[109]

Principles of Gas Anesthesia

Administration of gaseous anesthetics allows more precise control of the depth of anesthesia and facilitates artificial ventilation when the animal has been intubated. More complete descriptions of the principles of gas anesthesia can be found elsewhere.[110,111]

Administration of gaseous anesthetics generally requires specialized equipment such as an anesthetic machine which regulates the flow of anesthetic and oxygen. In addition, a scavenger system for waste anesthetic gases is necessary, since exposure to such gases poses a health risk to personnel. A variety of small animal anesthesia machines are commercially available.

Gaseous anesthetics may be administered through a face mask or a tube which has been passed into the trachea (endotracheal intubation). The face mask or endotracheal tube is connected to the anesthesia machine and the controls are set to deliver a specific concentration of anesthetic. Because rabbits may resist placement of the face mask or endotracheal tube, and because rabbits often hold their breath when exposed to gas anesthetics,[112] it is advisable to first sedate or lightly anesthetize the animal with an injectable agent.

If a face mask is used, the mask is gently fitted over the rabbit's face (Figure 27) and then secured with gauze tape tied around the back of the head once the animal is anesthetized. Placement of an endotracheal tube is discussed below. Alternatively, anesthesia can be induced by placing the animal in a gas anesthesia chamber, which connects to the anesthesia machine for delivery of the anesthetic gas. Once the rabbit has been lightly anesthetized, it is removed from the chamber and either the face mask or the endotracheal tube may be fitted. Both Bain's circuit and T-piece circuit arrangements can be used.

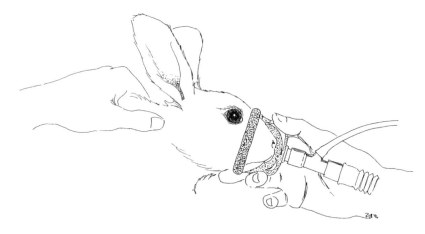

Fig. 27.   Anesthesia of rabbit by use of a face mask apparatus.

The flow rate depends on the size of the rabbit, but rates of 2.5 to 7.5 l/min are commonly used. More detailed descriptions of gas anesthesia in rabbits specifically can be found elsewhere.[93,110]

➤ *Equipment for Endotracheal Intubation*

1. Laryngoscope with size 1 Wisconsin blade if the rabbit is >3 kg or size 0 if <3 kg.

2. Endotracheal tube of 2.5 to 4.0 mm internal diameter. The length of the endotracheal tube should approximate the distance from the nose of the rabbit to a point several millimeters anterior to the thoracic inlet. Freezing the endotracheal tube prior to use will stiffen the tube and facilitate intubation.[113]

3. Water-soluble sterile lubricant.

4. Syringe to inflate endotracheal tube cuff.

5. Optional: lignocaine spray or 2% viscous lidocaine.

➤ *Procedure for Endotracheal Intubation*

1. The rabbit is lightly anesthetized with an injectable anesthetic.

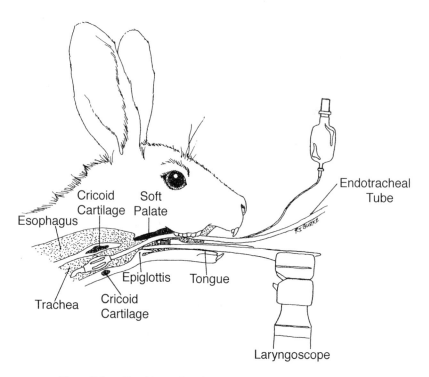

Fig. 28.   Endotracheal intubation of the rabbit.

2. The rabbit is placed on its sternum and the head and neck extended. Insertion of the endotracheal tube may be facilitated by positioning of the operator behind the rabbit and extending the head and neck of the rabbit upward.[114]

3. The laryngoscope is placed over the tongue and toward the back of the mouth to improve visualization of relevant structures.

4. Pressure is exerted on the tongue with the laryngoscope blade to open the epiglottis, which covers the opening to the trachea (Figure 28).

5. The larynx can be sprayed with lignocaine, or the endotracheal tube can be coated with viscous lidocaine to inhibit laryngeal spasm. A small amount of water–soluble sterile lubricant should be applied to the endotracheal tube to ease passage into the trachea.

6. The endotracheal tube is advanced into the trachea and the stylet removed, if a tube with a stylet is being used. Passage of the tube to the side of the laryngoscope channel will permit continued visualization of the larynx as the tube is advanced.[115]

7. The cuff of the tube is inflated, if a cuffed tube is being used.

8. Location of the tube within the trachea, rather than the esophagus, is confirmed by detecting passage of air through the tube as the rabbit breathes.

9. The external end of the tube is connected to the gas anesthesia machine.

10. Following anesthesia, the tube is removed gently once the rabbit exhibits the gagging reflex.

➤ *Alternative Procedures for Endotracheal Intubation*

- Blind intubation without a laryngoscope.[116-119]
- Percutaneous tracheal catheterization.[120]

Characteristics of Commonly Used Gas Anesthetics

Halothane

- Advantages: induces rapid anesthesia with good muscle relaxation.

- Disadvantages: may induce hypotension and cardiorespiratory depression.

- Usually administered at concentrations of 1.5 to 2.0%.

Methoxyflurane

- Advantages: produces effective anesthesia with good muscle relaxation.

- Disadvantages: recovery may be prolonged; potential renal and hepatic toxicity.

- Usually administered at concentrations of 0.4 to 1.0%.

Isoflurane

- Advantages: produces effective anesthesia, with less potential for untoward effects than halothane or methoxyflurane.

- Disadvantages: may induce respiratory depression and hypotension.

- Usually administered at concentrations of 1.0 to 3.0%.

Principles of Local Anesthesia

Local anesthesia involves desensitization of specific, defined sites, often involving the skin and subcutaneous tissues. Cutaneous biopsy and percutaneous blood sampling are examples of procedures that might involve local anesthesia. Injectable lidocaine and lignocaine-prilocaine cream are useful local anesthetics in the rabbit.

Lidocaine

- Produces excellent local tissue desensitization.

- Induction of local anesthesia involves infusion of a small volume by needle and syringe, usually 0.5 to 1.0 ml of 1% or 2% lidocaine, directly into or around the site of interest.

Lignocaine-Prilocaine Cream

- A commercial preparation of lignocaine-prilocaine cream (EMLA cream, Astra Pharmaceutical Products, Inc., Westborough, MA) can be used for skin desensitization prior to venipuncture of the rabbit marginal ear vein.[121]

- Cream is applied to the site and covered with an occlusive bandage. Desensitization takes place over approximately 1 hour.

Sedation and Tranquilization of Rabbits

Rabbits that resist handling or manipulation may need to be sedated or tranquilized. **Sedatives** produce drowsiness and reduced apprehension, while **tranquilizers** produce a calming effect without drowsiness. Most drugs used in this regard do

TABLE 12.   COMMON AGENTS FOR SEDATING/TRANQUILIZING RABBITS

Agent	Dosage/route	Reference(s)
Acepromazine	0.75–1.0 mg/kg, IM	122,123
Xylazine	3–9 mg/kg, IM	124
Diazepam	4–10 mg/kg, IM	125
Hypnosis	See "Manual Restraint" Chapter 5	126,127

not have a true analgesic effect but rather produce mental calming and increased tolerance for handling and other external stimuli. Agents frequently used for sedation or tranquilization of rabbits are listed in Table 12.

Analgesia

Any procedure which could potentially result in postprocedural pain should involve the use of an analgesic. Signs of pain in the rabbit include decreased food and water consumption, reluctance to move, vocalization manifested as a shrill cry, and subtle or obvious changes in attitude. Not all rabbits will exhibit overt indications of pain following painful procedures; thus analgesics should be routinely used for potentially painful procedures. Information for commonly used analgesics in rabbits is summarized in Table 13.

Perianesthetic Management

Because anesthesia can have a major physiologic impact, it is important that animals undergoing anesthesia receive proper supportive care before, during, and after the procedure.

➤ Care Prior to Anesthesia

Only healthy animals should be subjected to anesthesia for elective procedures. The following steps should be taken to assure that the animal is healthy prior to anesthesia.

- Evaluation of a blood sample for hematologic abnormalities.

- A brief physical examination including rectal temperature measurement and auscultation of the heart and lungs should be performed to assess the overall health status and to detect any signs of illness, particularly pasteurellosis or other respiratory disease.

TABLE 13. ANALGESICS COMMONLY USED IN RABBITS

Agent	Dosage/route	Degree of pain relieved, duration	Reference
Aspirin	100 mg/kg, PO in solution	Mild to moderate, 4 hours	128–130
Butorphanol tartarate	0.1–1.5 mg/kg, IV 1.0–7.5 mg/kg, SC or IM	Mild to moderate, 4 hours	131,132
Butorphanol + xylazine	5 mg/kg + 4 mg/kg, IM	Moderate, 2 hours	93
Buprenorphine	0.01–0.05 mg/kg, SC or IV	Severe, 6–12 hours	93,131
Morphine	2.5 mg/kg, SC	Severe, 2–4 hours	93,131

- The animal should be weighed so that the accurate dose of anesthetic may be determined.

Note: Although many species are fasted prior to anesthesia in order to minimize the potential for aspiration of vomited material during anesthesia, rabbits are unable to vomit; thus fasting is unnecessary.

- Subcutaneous fluids should be administered to prevent dehydration during the procedure. Sterile lactated Ringer's solution or 0.9% NaCl are appropriate fluids and can be easily administered subcutaneously at a dosage of 10 to 15 ml/kg of body weight for each anticipated hour of anesthesia.

- To minimize production of respiratory secretions, which may interfere with respiration, glycopyrrolate (0.1 mg/kg SC or IM) may be administered prior to anesthesia.[133]

Note: Although atropine is often used in other species to prevent excessive respiratory tract secretions, it is ineffective in many rabbits, due to rapid enzymatic degradation.[134]

➤ *Care during Anesthesia*

The anesthetized rabbit should be closely monitored to determine that the anesthetic depth is sufficient and to assure that the animal remains physiologically stable. In this regard, the following procedures should be followed:

1. The depth of anesthesia should be monitored prior to and during surgical or other experimental manipulations. Useful parameters to measure anesthetic depth include:

   - **Jaw tone.** Loss of resistance to opening of the mouth is one measure of anesthetic depth.

   - **Changes in heart and respiratory rates and character.** With some anesthetic regimens, increases in heart and respiratory rate and depth can indicate an insufficient level of pain insensation.

- **The pedal reflex.** With the leg extended, one toe is firmly pinched. In general, a surgical depth of anesthesia can be assumed if the rabbit does not vocalize nor attempt to withdraw the limb, although the reliability of this reflex can vary between rabbits.

- **The ear pinch.** Failure of the rabbit to arouse following pinching of the ear indicates a relatively deep level of anesthesia.

2. Adequacy of cardiovascular function should be monitored by means which might include:

   - Monitoring of heart and respiratory rate and character.

   - Electrocardiographic (ECG) pattern.

   - Blood gas levels and pH.

   - Blood pressure.[93]

3. Body temperature of the rabbit should be maintained to prevent hypothermia, particularly if undergoing experimental surgery involving exposure of a body cavity. Appropriate measures for this include:

   - Placement of the animal on blankets or heat blankets.

   - Administration of warm fluids.

   - Increasing the ambient temperature of the room.

➤ *Care Following Anesthesia*

It is critical that steps be taken to assure that the rabbit returns to a normal physiologic state following anesthesia. In this regard, the following measures should be taken:

- Monitoring of rectal temperature to assure that the rabbit does not become hypothermic. Hypothermic rabbits can be covered with blankets, placed on heat blankets, or placed at least 10 to 12 inches beneath a heat lamp until the body temperature has returned to normal.

- Monitoring of pulse and respiratory rate and character.

- Administration of supplemental fluids if needed.

> **Note:** A qualified veterinarian should be consulted if recovery from anesthesia is prolonged or otherwise abnormal.

aseptic surgery

All survival surgery in rabbits should involve aseptic techniques and preparation of the surgical site. Generally, this involves the following procedures and principles:

- Hair at surgical sites is shaved with an electric clippers.

- The skin is scrubbed thoroughly with an antiseptic.

- The surgical site is isolated with sterile surgical drapes.

- Sterile surgical instruments are used for all surgical procedures.

- Individuals performing surgery or directly assisting the surgeon wear sterile surgical gloves, gowns, caps, and face masks.

- Principles of aseptic surgery are described in greater detail elsewhere.[135-139]

> **Note:** Nonsurvival surgical procedures (those that do not involve recovery from anesthesia) need not be aseptic.

postsurgical management

Following surgery, animals should be monitored frequently until the experiment has been completed. Particular attention should be paid to the following:

FIG. 29. Rabbit wearing an elizabethan collar to prevent biting of sutures.

- Any change in behavior or appetite should be noted. Changes may indicate that the animal is in pain or experiencing other complications.

- The edges of the surgical incision should remain neatly apposed, following surgery. Surgical sites must be examined to assure that rabbits do not remove sutures. Equipment such as **Elizabethan** (Figure 29) or **cervical collars** (Figure 30) are useful to restrict the ability of the rabbit to bite sutures, but do not interfere with the rabbit's ability to eat. **Tissue glue** is a useful alternative to sutures in rabbits, since the lack of a suture often diminishes the interest of the animal in the incision site.

- Signs of infection at the surgical site, in the days and weeks after surgery. Signs of infection include:

  - Thick, white, yellow, or green discharge.

  - Abnormal warmth or redness.

  - Elevated body temperature.

  Any discharge should be cultured for bacteria, and antibiotic therapy begun under the guidance of a qualified veterinarian.

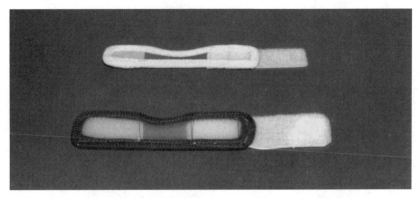

FIG. 30.   Cervical collars.

euthanasia

Euthanasia of animals should be conducted in a humane and professional manner. The specific method chosen should produce a quick, painless death. Individuals euthanizing rabbits should be properly trained in the euthanasia technique to be used. In general, animals should be euthanized out of the sensory range of other animals. Death should always be confirmed by stethoscopic auscultation for absence of a heartbeat, followed by cutting of the diaphragm to prevent unexpected resuscitation. Specific methods of euthanasia for rabbits are summarized in Table 14. A complete summary of recommendations for euthanasia can be found in the 1993 "Report of the American Veterinary Medical Association Panel on Euthanasia."[140]

TABLE 14.   COMMON METHODS FOR EUTHANASIA OF RABBITS

Method	Route/technique	Comments
Pentobarbital overdose	150 mg/kg IV or IP	Use of pentobarbital requires Drug Enforcement Agency Registration
Carbon dioxide overdose	Inhalation of 60 to 70% CO_2 in an appropriate chamber	Compressed CO_2, in a cylinder, NOT dry ice, should be used as CO_2 source
Cervical dislocation	Neck is hyperextended and twisted, separating first cervical vertebra	Do NOT perform on unanesthetized rabbits, nor those weighing >1 kg

experimental methodology

Rabbits have been used in a myriad different ways in research. Although it is impossible to cover all the techniques which have been used, some of the more common methods and principles for handling rabbits during experimentation are presented here. It is important that personnel performing experimental manipulations on animals are properly trained in those procedures.

restraint

It is critical that rabbits be properly restrained to ensure safety of both the animal and the handler. The relative strength of the hind limbs of the rabbit to the fragility of the skeleton may translate into fracture of the lumbar spine if the rabbit's weight is not supported properly. Rabbits should never be grasped and carried by the ears. In addition, improperly held rabbits may inflict scratches on the handler with their rear feet.

> **Note:** The rear quarters of the rabbit must be supported during restraint.

In general, rabbits should be restrained for only the minimum amount of time needed to perform any experimental manipulation. Restraint methods include manual restraint or use of restraint devices.

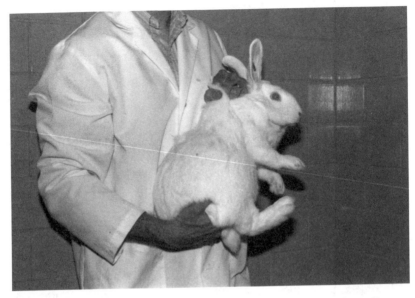

Fig. 31. Holding a rabbit by the scruff of the neck with one hand while supporting the rear quarters with the other hand.

Manual Restraint

The following principles should be considered regarding manual restraint:

- Rabbits can be picked up or lifted out of the cage by gently, but firmly, grasping the scruff of the neck with one hand and supporting the weight of the animal with the other hand. Never grab or lift a rabbit by its ears.

- The weight of the animal can be supported with one hand underneath the rear quarters of the rabbit (Figure 31) or by the handler holding the rabbit close to the body, with the rabbit's head tucked into the crook of the weight-supporting arm (Figure 32).

- Some aggressive rabbits may resist grasping and may even charge toward the handler. In such instances, a cloth towel gently thrown over the rabbit's face and body often tempers the rabbit's agression and facilitates handling.

- Restraint of rabbits following induction of a hypnotic-like state has also been described.[126] Rabbit **"hypnosis"**

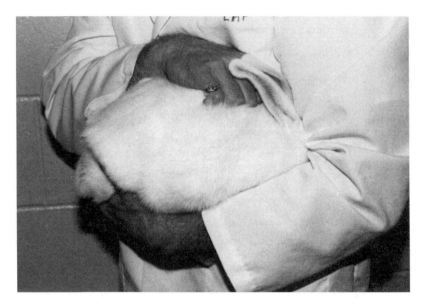

FIG. 32.   Holding a rabbit by the scruff of the neck and supporting the body with the other arm.

is performed by placing the rabbit on its back and placing the ears of the rabbit over the eyes, while the chin is firmly flexed against the neck and the body stretched out. After several minutes, the rear legs are released, and the thorax and abdomen are gently stroked, resulting in a hypnotic trance. The resulting length and degree of immobilization are inconsistent between rabbits, but may permit short procedures lasting no more than several minutes to be performed. Wide variations between rabbits with respect to analgesic effect argue against use of this technique for invasive, potentially painful procedures.[127]

Mechanical Restraint Devices

A variety of restraint devices have been developed for rabbits, some of which are available commercially.

* Most restraint devices are designed to hold the rabbit securely while allowing access to sites for administration of compounds or sampling (Figure 33).

FIG. 33. Rabbit in a restrainer.

- It is important that such equipment be sanitized following each use, since infectious disease can be transmitted
  between rabbits via equipment.

---

sampling techniques

Rabbits are frequently used in studies which require sampling
of blood, urine, cerebrospinal fluid, or bone marrow. Common
methods for obtaining such samples are presented below.

Blood

Frequently, large–volume blood samples are required of rabbits, particularly those used for polyclonal antibody production.
General principles for blood sampling include:

- **Volume of sample.** If too much blood is removed at a
  single time, the animal may experience hypovolemic
  shock and could die. If blood sampling is too frequent,
  the animal may become anemic. In general, 10% of the
  circulating blood volume can be removed every 3 to 4
  weeks, or 1.0% at more frequent intervals of 24 hours

or more, with minimal adverse effect to the rabbit.[141] The circulating blood volume is approximately 55 to 70 ml/kg of body weight; thus 6 to 7 ml/kg of body weight can be safely removed every 3 to 4 weeks. It is best to alternate sample sites each time blood is withdrawn.

- **Sampling vials.** Samples needed for evaluation of whole blood are usually collected in vials containing an anti-coagulant such as EDTA. Samples needed for collection of plasma are collected in vials containing heparin, cit-rate, or potassium oxalate, depending on the experimen-tal use. Samples for harvest of serum are collected in vials containing no anticoagulant. Blood sampled for processing to plasma or serum can be centrifuged at approximately 800 to 1000 × g for 10 to 15 minutes and the liquid fraction harvested as the sample. Alternatively, the liquid fraction will generally separate from the clotted cellular fraction if the sample is left at room temperature for approximately 30 minutes.

Blood sampling sites and methods include:

- **Percutaneous sampling**. General principles include:

  1. Usually involves percutaneous vascular access by means of a syringe and sterile hypodermic needle.

  2. The bore of the needle should ideally be just less than the diameter of the vessel, to ensure rapid blood withdrawal. Useful sizes include 21-25 gauge needles.

Note: Hypodermic needle diameter increases with decreasing gauge number.

---

  3. Disposable needles and syringes are commonly used, and should be disposed of in an appropriate, puncture-resistant container designated for sharp objects. To minimize the risk of needle-stick injury to personnel, the needle, still attached to the syringe, should be disposed of without replacing the needle cap over the needle or bending the needle.[56]

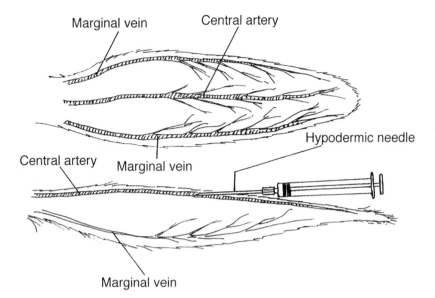

FIG. 34. Location of the ear vessels from a dorsal view (top) and from a lateral view, indicating relative orientation of the needle during sampling of blood from the central artery (bottom).

sites for percutaneous blood sampling

Several sites can be used for obtaining blood samples percutaneously, including:

➤ *Marginal Ear Vein or Central Ear Artery (Figure 34)*

These are the most commonly used sites for sampling of blood from rabbits. The following points should be considered:

1. The artery is preferred for larger volume samples.

2. Sampling of blood from the vein should be performed as close to the base of the ear as possible, whereas sampling from the artery should be performed nearer to the tip of the ear. If initial attempts at venipuncture at these sites are unsuccessful and result in hematoma formation, additional attempts can be made distally toward the ear tip for the vein and proximally toward the base for the artery.

3. The vessels can be dilated by wrapping the ear for several minutes in a towel soaked with warm water

or by vigorously rubbing or massaging the ear for 30 to 60 seconds. **Xylene** can be topically applied for vasodilation; however it is irritating and can result in dermatitis if the ear is not thoroughly washed afterwards. Topical **d-limonene**[142] and **citrus oil**[143] are also vasodilatory. Application of 2 ml of a 40% *d-*limonene solution in 95% ethanol stimulates vasodilation within several minutes. Alternatively, the sedative **acepromazine maleate** not only calms difficult rabbits, but is also vasodilatory at IM dosages of 1 to 5 mg/kg body weight.[141]

➤ *Jugular Vein*

1. The jugular vein cannot be externally visualized in rabbits, thus jugular venipuncture is made following location of the vessel based on anatomic landmarks and palpation.

2. The rabbit is laid on its back and the neck extended.

3. The vessel is palpated in the neck after the technician has occluded the vessel at the apex of the sternum, by pressing with the thumb at the thoracic inlet.[144] Clipping of the fur in the region of the jugular vein may facilitate identification of the vessel.

➤ *Cardiac Puncture*

Cardiac puncture should be used for obtaining large volumes of blood when sampling is to be immediately followed by euthanasia. Procedures for cardiac puncture are:

1. The rabbit is anesthetized.

2. The rabbit is placed on its back.

3. A long, large–bore needle, such as an 18-gauge, 1 1/2 inch needle, is inserted cranially and dorsally at a 30° angle just caudal and left to the xiphoid process (Figure 35).

4. Alternatively, the anesthetized rabbit may be placed on its right side and the needle inserted between the ribs at the point of maximal palpable heartbeat intensity and directed toward the sternum. Due to laevoversion of the heart in some rabbits, cardiac

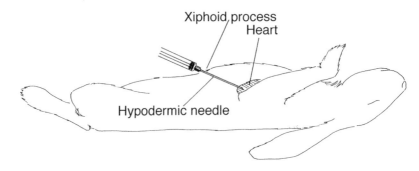

Xiphoid process
Heart
Hypodermic needle

Fig. 35. Cardiac puncture in the rabbit.

puncture may be facilitated in those animals by placement of the rabbit in left lateral recumbency.

Note: Cardiac puncture is best performed only on animals immediately prior to euthanasia, due to the risk of cardiac tamponade and laceration of pleural viscera.

➤ *Method for Percutaneous Blood Sampling*

The following steps can be used for percutaneous blood sampling:

1. The hair around the site is shaved and the site swabbed with an antiseptic such as rubbing alcohol to minimize the chance of introducing skin-associated bacteria into the blood stream.

2. The location of the blood vessel is identified.

3. The plunger of the syringe is slightly pulled back before use, thereby breaking the air lock and allowing blood flow to begin more easily.

4. The needle is directed with the beveled edge up at a slight angle into the vessel, and the plunger of the syringe is gently and slowly pulled back as blood fills the shaft of the syringe. If the plunger is pulled too aggressively, the vessel may collapse and blood flow will cease.

5. Gentle manipulation such as slight changes in the orientation of the needle may improve collection if

blood flow slows or ceases. Alternatively, hematoma formation in the tissues around the withdrawal site or the presence of clotted blood in the needle may necessitate changing withdrawal site or replacement of the needle, respectively.

6. Once the sample has been obtained and the needle withdrawn, firm pressure with a gauze pad should be maintained at the sampling site for several minutes until bleeding ceases.

➤ *Vascular Catheterization*

In situations requiring repeated blood sampling from rabbits, implantation of a vascular catheter may be merited.

1. Techniques have been described for catheterization of various sites including:

   • The marginal ear vein[145-147]

   • The central ear artery[148]

   • The right anterior vena cava[149]

   • The right atrium of the heart[150]

2. **Subcutaneous vascular access ports** allow chronic vascular access without an exposed catheter.[151]

3. Although vascular catheters can be maintained for periods of months or even longer, catheter tract bacterial infections and bacteremia are common complications[152,153] which generally depend upon removal of the catheter for successful resolution.[154] In this regard, use of **aseptic technique** in catheter implantation, and **frequent cleaning and flushing** of catheters are fundamental to successful long–term catheter maintenance.[150] Patency of catheters can be maintained by flushing 0.5 to 1 ml of sterile heparinized saline (100 IU/ml) through the catheter every other day.

4. Secondary renal and peripheral vascular abnormalities can result from long-term vascular catheterization in rabbits.[155]

> **Note:** Chronic vascular catheterization facilitates repeated blood sampling and intravenous compound administration but requires regular maintenance to avoid associated health problems.

Urine

The method of urine collection depends on the use of the sample and the need for sample free from contamination with bacteria, feces, or other debris. Urine should be collected in a clean, dry container and stored under refrigeration if the sample is not to be used within several hours.

1. **If sample contamination is not a significant concern**, urine can be sampled directly from a pan placed beneath the rabbit if it is housed on a wire grid. Alternatively, the urinary bladder may be palpated in the caudal abdomen and gentle pressure applied to express urine from the bladder.

2. **Relatively pure** urine samples may be collected by **catheterization** of the urinary bladder. To perform this technique, a lubricated, sterile, 9 French flexible catheter is passed through the urethra and into the urinary bladder. Preferably, rabbits are first sedated or lightly anesthetized. In male rabbits, the animal is restrained in an orientation, such as a sitting position, that allows access to the penis and the catheter is directed into the urethra and downwards (Figure 36). Treatment of male rabbits with the sedative acepromazine (5 mg/kg, IM) frequently stimulates temporary protrusion of the penis from the prepuce[156] and could facilitate urinary catheterization of bucks. In female rabbits, the urethral opening is located on the floor of the vagina and must be gently reflected dorsally with the catheter tip to allow passage into the urethra (Figure 37). Once the catheter has entered the urinary bladder, urine should flow through the catheter following gentle manual compression of the caudal abdominal region and under gentle aspiration pressure from an attached syringe.

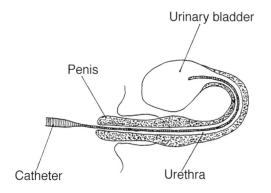

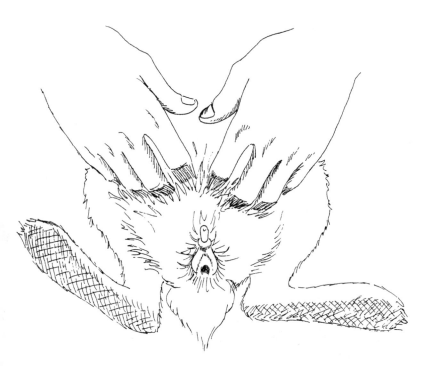

Fig. 36. Extrusion of the penis and palpation of the urinary bladder (bottom) and orientation of the catheter in the urinary tract during catheterization of the bladder of the male rabbit (top).

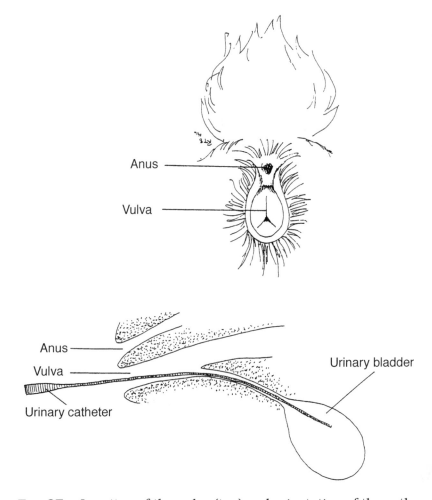

Fig. 37. Location of the vulva (top) and orientation of the catheter in the urinary tract during catheterization of the female rabbit (bottom).

3. **Relatively pure** urine samples may also be obtained by **cystocentesis**, a procedure in which a hypodermic needle attached to a syringe is passed through the abdominal wall and into the urinary bladder (Figure 38). To perform cystocentesis, the rabbit is placed in dorsal recumbency, and the caudal ventral abdomen prepared as for aseptic surgery. Preferably, the rabbit is first sedated or lightly anesthetized. Following palpation of the abdomen to locate the urinary bladder, a 20- to 22-gauge hypodermic

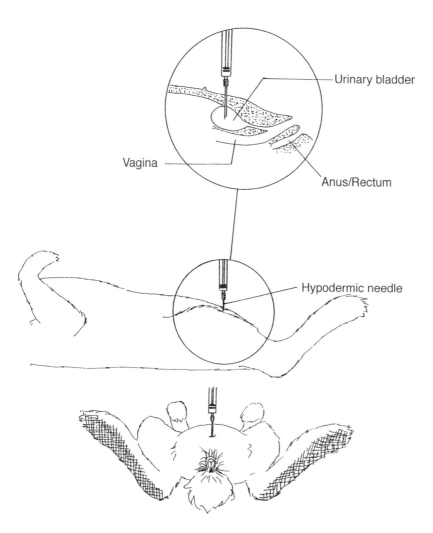

FIG. 38. Position of the hypodermic needle from a caudal view (bottom), lateral view (middle), and orientation of the needle during cystocentesis of the female rabbit (top).

needle is inserted through the skin and into the bladder. Urine can then be sampled under gentle aspiration pressure from an attached syringe. The urinary bladder may be difficult to palpate if the rabbit has recently urinated.

4. Urine to be sampled **over extended periods of time,** can be collected by means of a **metabolism cage.**

Such cages employ a mechanism which diverts the urine into a collection receptacle separate from the feces.

Cerebrospinal Fluid

Collection of cerebrospinal fluid (CSF) from rabbits usually involves:

- Insertion of catheters or spinal needles.

- Aseptic techniques including aseptic preparation of the sampling site when needles or catheters are inserted.

- Anesthesia of the rabbit, since significant damage to structures of the nervous system can occur if the rabbit is not completely immobile during insertion of needles or catheters for CSF collection.

Specific sites and techniques for CSF collection include:

- **The cerebromedullary cistern** — an enlargement of the **fourth ventricle** at the atlantooccipital junction. Single CSF samples of 1.5 to 2 ml volume are relatively easy to obtain by needle and syringe from this location.[157]

➤ *Procedure for sampling CSF from the cerebromedullary cistern:*

1. The rabbit is anesthetized.

2. The dorsal cervical area of the rabbit is prepared as for aseptic surgery.

3. The rabbit is placed in lateral recumbency. With a firm grasp and placement of the thumb on the bony occipital protuberance, the head is flexed toward the chest while taking care not to obstruct the airway (Figure 39).

4. A 22-gauge, 1.5-in. spinal needle is slowly inserted approximately 2 mm caudal to the occipital protuberance and advanced slowly in the direction of the rabbit is mouth. It should be remembered that

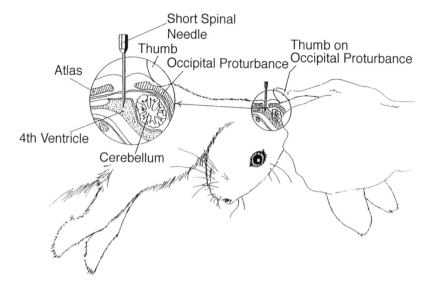

FIG. 39. Sampling of cerebrospinal fluid from the rabbit. The needle passes into the 4th ventricle lateral to the spinal cord.

the epidural space of the rabbit is only 0.75 to 2.5 cm beneath the surface of the skin[158,159] in the lumbosacral region and likely is at a similar depth in the cervical region. The needle is kept parallel with the table surface, with slight rotations of the rabbit's head to maintain the alignment as needed.

5. A slight decrease in resistance may be felt as the needle enters the cistern. The stylet is then removed and CSF flow is usually apparent at the hub of the needle.

6. CSF can then be harvested by **gentle** suction from an attached syringe. Small blood vessels are often lacerated during needle insertion; thus contamination of the sample with blood may occur with this technique.

• **The cisterna magna.** Catheterization of this site permits repeated, chronic sampling of CSF. The advantages of such procedures are that CSF can be sampled in a nonanesthetized animal and that CSF samples obtained in this way are generally free of blood contamination.

Methods for chonic CSF sampling from the cisterna magna include:

1. Implantation of a catheter with sampling through an implanted subcutaneous access port.[160]

2. Implantation of a catheter through a drilled hole in the skull. CSF is sampled through this catheter which is anchored externally on the top of the head.[161]

- **Lumbar subarachnoid space.** This site can be used to sample CSF through an implanted catheter which is tunneled subcutaneously to exit externally through the skin.[162]

- **Third ventricle.** Placement of a permanent stainless steel cannula into the third ventricle through a drilled hole in the skull has been used as a means to collect multiple CSF samples from the unanesthetized rabbit.[163]

Bone Marrow

Sampling of bone marrow from rabbits should be performed with the animal properly anesthetized. Aseptic technique, including site preparation, should be followed. Sampling sites include the:

- **Proximal end of the tibia** by means of percutaneous aspiration using a 15-gauge needle.[164]

- **Humerus** by means of percutaneous aspiration using a Rosenthal pediatric needle.[165]

- **Femur** by means of percutaneous aspiration using a Rosenthal pediatric needle.[165] Alternatively, marrow can be obtained through a surgical approach involving resection of the femur.[166] While this technique allows direct access to sterile marrow, the weakened femur is at risk of postsurgical fracture due to the weight bearing function of the rear limb.

---

compound administration techniques

A variety of routes exist for administration of both test compounds and medications to rabbits. As with sampling techniques, it is important that the rabbit be securely restrained. When possible, rabbits should be sedated or lightly anesthetized to minimize any potential pain or distress. Routes of compound administration include:

Intravascular

Intravascular administration of compounds results in quick delivery to target tissues. Unless specifically required by experimental protocol, substances given intravascularly should be administered slowly, so that the consequences of an unexpected adverse reaction can be minimized. The following points for intravascular compound administration should be considered:

Common sites: marginal ear vein or jugular vein. The high pressure within the central ear artery precludes practical use of that vessel for compound administration by needle and syringe.

➤ *Technique*

Liquid compounds may be administered by needle and syringe using essentially the same approach as for blood sampling from these sites. The technique is as follows:

1. Before injection of the compound, the syringe plunger should be pulled back slightly to confirm that the needle is indeed within the vessel. Intravascular location is confirmed by the presence of blood in the hub of the needle and the tip of the syringe shaft. Formation of a bleb or blister within the skin around the vessel, as compound is being injected, indicates that the needle is not within the blood vessel.

2. The compound is administered slowly. If the rabbit develops complications such as respiratory distress, additional compound should not be administered.

Chronic intravascular administration of compounds may be best performed through implanted catheters. Sites and techniques for catheters include:

1. **Posterior facial vein**.[145]

2. **Marginal ear vein** using polyethylene tubing (inner diameter = 0.011 in., outer diameter = 0.024 in.) attached

to a 2.8 French vascular access port affixed to the ear with cyanoacrylate adhesive.[146]

3. The **jugular vein** can be used for chronic serial infusion by means of a surgically implanted vascular access port.[151]

Continuous infusion in mobile, unrestrained rabbits can be performed using a swivel–tether system and catheterization of either the common carotid artery[149] or the marginal ear vein.[167,168]

Intramuscular (IM)
Sites: the large **gluteal** muscles on the back of either of the rear legs (Figure 40) or the **perilumbar** musculature of the back (Figure 41).

➤ Technique

1. Depending upon the viscosity of the compound, the smallest bore needle possible should be used. Generally, 22- to 26-gauge needles are useful for this application.

2. It is important that nerves and blood vessels be avoided when giving IM injections. The muscle mass should be firmly grasped and immobilized with one hand, while the other hand manipulates the syringe.

3. The skin is swabbed with alcohol and needle is inserted, bevel up, through the skin and into the muscle.

4. The syringe plunger is gently pulled back to ensure that the needle is not within a blood vessel. If intravascular, a small amount of blood may appear in the syringe or the hub of the needle, and the needle should be repositioned.

5. In general, volumes no greater than 0.5 ml should be injected into a single IM site.

Subcutaneous (SC)
Site: dorsum (top) of the neck and back. Subcutaneous administration of compounds by needle and syringe is easily

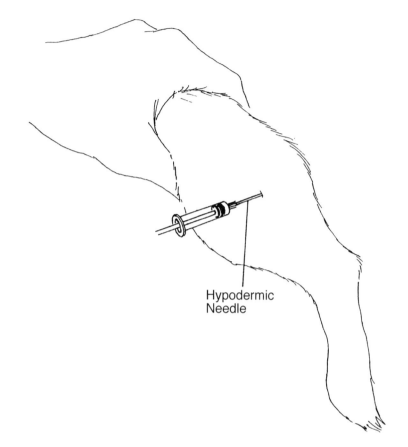

Hypodermic
Needle

FIG. 40. Administration of an intramuscular injection in the rear leg of the rabbit.

accomplished in rabbits due to the pliable skin and large sub-cutaneous space on the dorsum (top) of the neck and back of the rabbit.

➤ *Technique*

1. The smallest bore needle possible should be used depending upon the compound viscosity. Generally, 21-gauge or greater (smaller diameter) needles are useful.

2. A fold of skin on the neck or back is lifted and swabbed with alcohol.

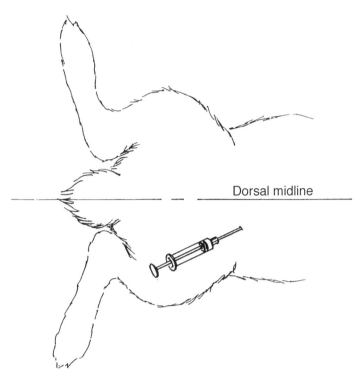

Dorsal midline

FIG. 41. Administration of an injection into the perilumbar muscles of the back in the rabbit.

3. The needle, with attached syringe, is inserted at a right angle to the skin fold (Figure 42). Care must be taken so that the needle is not passed completely through the skin fold to exit the other side.

4. Relatively large volumes (generally 10 to 20 ml per subcutaneous site) can be injected subcutaneously, although excessive volumes which greatly distend the skin should be avoided. Subcutaneous injection of antigens for polyclonal antibody production usually utilize volumes of 0.10 to 0.50 ml per site.[169]

Intradermal

Intradermal injection is frequently used for immunization with specific antigens in the course of polyclonal antibody production.

Site: The skin on the dorsal abdomen and thorax (the back).

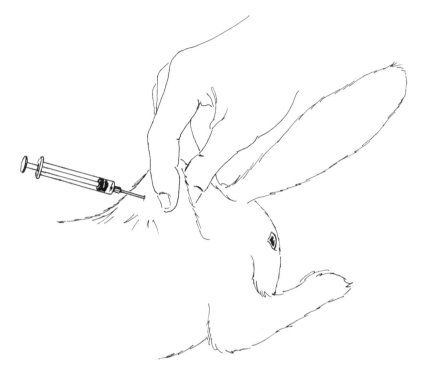

FIG. 42. Subcutaneous injection in the rabbit.

➤ *Technique*

1. The hair is shaved to adequately view the injection site.

2. The skin is swabbed with alcohol, and a small bore needle (25- to 27-gauge), bevel side up, attached to a tuberculin syringe is used to penetrate into the dermis while the skin is held taut (Figure 43). Alternatively, the needle may be inserted subcutaneously and then directed up toward an intradermal location.

3. A small blister-like bleb forms within the skin on injection, confirming the intradermal location.

4. Small volumes of approximately 0.05 to 0.1 ml per site may be administered in this manner.

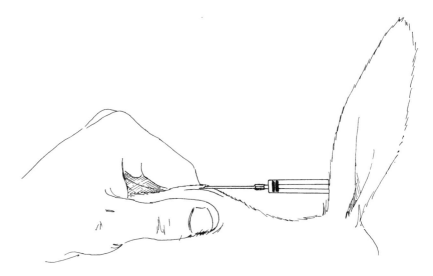

FIG. 43. Intradermal injection in the rabbit.

Intraperitoneal (IP)

Compounds are occasionally administered to rabbits intraperitoneally; however the risk of accidental puncture of and injection into abdominal organs exists.

Site: Lower (caudal) right abdominal quadrant.

➤ *Technique*

1. Normally a long needle (1 inch or greater) is used.

2. The rabbit is placed on its back, which may require sedation in some rabbits. Alternatively, a lone technician may restrain an unsedated rabbit by securing the head and front quarters of the rabbit with his or her legs and extending the lower portion of the rabbit's body with one hand (Figure 44).

3. The rabbit's hindquarters are held at a 30° to 45° angle to the horizontal, if the rabbit is on its back.

4. The needle is inserted into the lower right abdominal quadrant just lateral to the midline and directed at an approximately 45° angle to the body wall.

5. The syringe plunger is gently pulled prior to injection to ensure that neither viscera nor blood vessels have

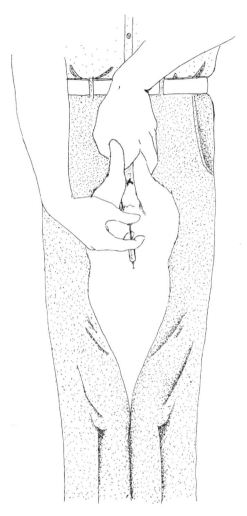

FIG. 44. Restraint and intraperitoneal injection in the rabbit by a lone technician.

been penetrated. For example, aspiration of yellow fluid implies that the needle has penetrated the urinary bladder, while green fluid suggests that the intestinal tract has been penetrated. Unexpected contamination by such materials necessitates discarding the compound to be given and obtaining a fresh sample.

Implantable Osmotic Pumps

Osmotic pumps are small, self-contained devices which are designed to deliver substances at a specified rate under the force of osmotic pressure. The pump can be implanted surgically into subcutaneous sites or into the abdominal cavity.

Oral

There are several methods for oral administration of compounds, including:

- **Incorporation of compound into the drinking water or food** can be used when oral delivery of compounds need not be exactly precise.

- **Administration by syringe**. Small volumes may be administered using a syringe by placing the tip of the syringe at the corner of the rabbit's mouth and slowly injecting the material.

- **Capsule administration**.[170,171]

- **Surgical pharyngotomy**.[172]

- **Oral gavage** is useful for delivery of specific amounts of compound. Oral gavage is performed in the following manner:

➤ *Technique*

1. Fractious rabbits may need to be sedated for oral gavage.

2. The total length of tube to be inserted can be estimated as the length from the mouth to the last rib and should be marked on the tube before insertion is begun.

3. A speculum is placed in the rabbit's mouth to prevent chewing of the tube. A small block of wood with a hole drilled in the middle to allow passage of the tube is sufficient.

4. The tube (usually an 8 French infant feeding tube) is lightly lubricated with petroleum jelly.

5. The tube is passed through the speculum and back to the pharynx. When the rabbit demonstrates the

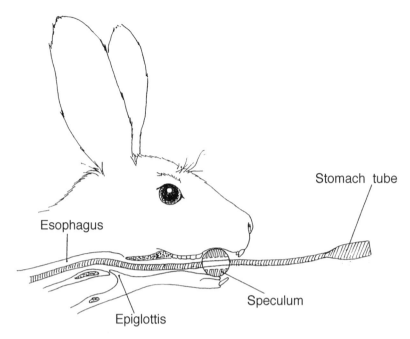

FIG. 45.   Orientation of the stomach tube for oral gavage in the rabbit.

gag reflex, the tube is advanced into the esophagus and on into the stomach (Figure 45).

6. The location of the tube in the stomach must be confirmed to avoid accidental administration of compound to the respiratory tract. This is most easily accomplished by examining for a lack of air passage through the tube as the rabbit breathes.

7. Compound is slowly administered by a syringe attached to the stomach tube. Volumes as large as 10 ml/kg can be safely administered in this way. It is advisable to subsequently administer a small volume of water to rinse any residual compound into the stomach.

8. After administration, the tube is kinked, to prevent flow of residual material in the tube into the respiratory tree as it passes through the pharynx, and the tube is slowly withdrawn.

polyclonal antibody production

Rabbits are commonly used for production of antibody to a variety of antigens. Typically, the antigen of interest is administered to the rabbit several times and the antibody-containing serum is later harvested. Because antibody to more than a single epitope of the antigen may be generated, antibody obtained in this way is referred to as **polyclonal**.

➤ *Key Points*

1. The exact amount to be administered varies with antigen; however 50 to 1000 µg of antigen per immunization is commonly used.[173]

2. Following initial immunization, booster injections are typically administered every 4 to 6 weeks, although the optimal time period may vary widely between antigens. It is preferable not to give booster injections until the serum antibody response begins to wane.

3. Because some rabbits may mount only a weak antibody response to the administered antigen, it is useful to immunize at least two or three rabbits for each antigen.

4. Booster immunizations can be given using essentially the same procedures as for the initial immunization. If Freund's complete adjuvant was used for the initial immunization, it should not be used for subsequent booster immunizations. Rather, Freunds' incomplete adjuvant could be used for booster immunizations (see following section on adjuvants).

5. Once immunized, rabbits may produce sufficient levels of antibody, often following a booster, to warrant maintaining them for periods of time lasting up to several years. For this reason, it is advisable to use specific pathogen-free rabbits to minimize the risk of loss of valuable rabbits to infectious disease.

6. Common routes of antigen administration are by injection intradermally, subcutaneously, or intramuscularly. Intravenous administration can also be used; however antigens given by this route may

induce anaphylatic shock, particularly upon booster administration.

Adjuvants

Adjuvants are nonspecifc stimulants of the immune system which are coadministered with antigens to promote a stronger antibody response to that antigen. The optimal immunization regimen and adjuvant appears to vary with individual antigens. Substances commonly utilized as adjuvants for polyclonal antibody production include:

1. **Freund's complete adjuvant** (CFA), which is the most commonly employed adjuvant and consists of a water-in-oil emulsion containing killed *Mycobacterium tuberculosis*. Antigen is thoroughly mixed with an equal volume of CFA to form an emulsion. The emulsion should be thick enough that it does not disperse when a drop of it is placed on the surface of a saline solution. CFA results in a chronic granulomatous response at the injection site,[174] which in the case of intradermal injections is often manifested as swelling and ulceration of the skin (Figure 46). When CFA is used as an adjuvant, only very small volumes should be injected per intradermal site. A suggested maximal volume is 0.05 ml.[169] CFA should be used only for the first immunization, since severe hypersensitivity reactions may result if CFA is used for booster immunizations. Instead, booster immunizations may contain no adjuvant or **Freund's incomplete adjuvant** (FIA), which does not contain mycobacterial cells. For many antigens, use of CFA in the initial injection and FIA in booster injections consistently enhances the antibody response.[175,176]

2. **Oil-in-water adjuvants**, which produce miminal inflammatory lesions and are an alternative to CFA. The **Ribi Adjuvant System** (Ribi Immunochem Research, Inc., Hamilton, MT) is an adjuvant containing fractions of the mycobacterial cell wall. In contrast, **TiterMax** (CytRx, Inc., Norcross, GA) is an oil-in-water emulsion containing a synthetic block copolymer. The methods for use are similar to those for CFA.

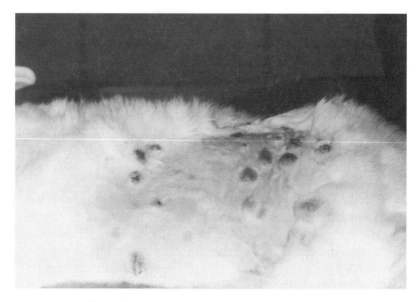

FIG. 46. Inflamed, ulcerated lesions at sites of intradermal injection of Freund's complete adjuvant on the back of a rabbit.

3. **Aluminum hydroxide** (alum), which can be used as an adjuvant by allowing adsorption of antigens onto an aluminum salt.[177,178] Use of alum will not result in inflammatory lesions such as those seen following use of CFA.

Collection of Antibody

The optimal time to collect serum for antibody harvest varies greatly with antigen, adjuvant, and route of immunization. The level of antigen-specific antibody can be measured by assaying a small serum sample 7 to 14 days after immunization and harvesting a larger volume of serum when a significant rise in antibody is noted. In general, one can expect high antibody levels approximately 10 days after each boost. Blood sampling procedures are described earlier in this chapter.

Subcutaneous Chamber Method

Polyclonal antibody may also be produced and collected by means of a surgically implanted subcutaneous chamber.[179-181] The technique involves subcutaneous implantation of a perforated, hollow, plastic golf ball which becomes encased within the tissue after several weeks. Antigen is percutaneously injected directly into the chamber through the perforations of the golf ball. Likewise, fluid rich in antigen-specific antibody is harvested percutaneously. The method allows collection of 12 to 22 ml of such fluid weekly. Adjuvants are not needed with this method.

safety testing procedures

Rabbits are frequently used for the evaluation of the toxicity and safety of compounds and formulations, many of which are intended for commercial distribution. Tests examining dermal and ocular irritation, teratogenicity, and pyrogenicity all commonly employ New Zealand White rabbits. Conduct of these tests is specified by various regulatory agencies of the U.S. federal government as well as regulatory agencies abroad.[182]

Dermal Irritation

Various modifications of procedures for testing of dermal irritancy are summarized elsewhere.[183] In general, the following procedures may be employed for conduct of the dermal irritation test:

➤ *Procedures*

1. The hair from a portion of the back of the rabbit is clipped, being sure to consistently use the same anatomic location in all rabbits tested.

2. After 24 hours, the test material is applied to the clipped area. If the test material is a liquid, 0.5 ml of the undiluted sample is applied, while 0.5 g is applied if the sample is a solid. Samples that are pH ≤ 2 or ≥ 11.5 are assumed to be dermal irritants and do not need to be tested.

3. Following sample application, the site is covered with a gauze pad fastened to the skin with tape.

4. After an exposure period of 4 to 24 h, as determined by the relevant regulatory agency, the sample is washed off with water or an appropriate solvent.

5. After 30 minutes, the skin is scored for redness and swelling according to a standard scale.[184] The skin site is also scored at 24, 48, and 72 h after sample removal.

6. The site should be monitored for up to 21 days after sample removal to determine the reversibility of changes in the skin.

Ocular Irritation

Various methods for assessing ocular irritancy have been summarized elsewhere.[185] In general, the following procedures may be followed to conduct the ocular irritation assay:

➤ *Procedures*

1. An ocular exam is conducted on all of the rabbits prior to testing to rule out preexisting ocular abnormalities.

2. One eye per rabbit is treated with test sample, while the contralateral eye serves as an untreated control. Liquid samples (0.1 ml) and solid samples (0.1 ml or ≤100 mg) are tested by placing the sample into the small pocket formed when the lower eyelid is gently pulled out. The eyelids are then held together over the eye for 1 second and then released.

3. The eyes are evaluated at time points specified by and varying with different regulatory agencies. Time points for evaluation often include 1, 24, 48, and 72 h after administration of test compound. The eyes are scored as "positive" if changes such as corneal ulceration or opacity, or inflammation or other changes are noted in the conjunctiva or other ocular tissues. If four of the six rabbits required per sample are scored positive, then the test is considered positive. If only two or three out of the six are scored positive, the test should be repeated. When repeated, three positive animals constitue a positive test and two positive animals necessitate a third test, which is positive if any rabbits are scored positive.

Teratogenicity

Studies designed to test the effect of compounds on the developing rabbit embryo generally involve compound administration to pregnant does from days 6 or 7 through day 18 of gestation. This may involve either maintaining breeding stock of rabbits or purchase of timed-bred rabbits from commercial vendors so that rabbits at precise stages of gestation are available. Procedures for teratogenicity testing are presented in detail elsewhere.[186] In general, the following procedures may be followed for conduct of teratogenicity testing:

➤ Procedures

1. Usually 3 test doses of each compound are evaluated, with 12 to 15 pregnant rabbits tested per dose. Generally, the highest dose tested should result in some degree of maternal toxicity, while the lowest dose should result in no maternal toxicity. Maternal toxicity may be characterized by weight loss, anorexia, lethargy or other clinical signs, or by death of the doe.

2. Test samples are administered by the likely route of human exposure.

3. After dosing, does are monitored daily for body weight, food and water consumption, and other clinical signs.

4. One to two days before expected parturition, the doe is humanely euthanized and the uterus examined for live and dead fetuses and resorption sites.

5. Fetuses are examined for gross and histologic abnormalities of the viscera and skeleton.

Pyrogen Testing

Preparations of pharmaceutical compounds are routinely evaluated for contamination with bacterial endotoxin or other pyrogens prior to distribution for use.[187] In general, pyrogen testing generally consists of the following procedures:

1. The test sample is intravenously administered to a rabbit for which a baseline rectal temperature has previously been established.

2. Rectal temperature is monitored hourly for 3 hours.

3. The sample is considered acceptable (free of endotoxin) if the rectal temperature does not exceed specified levels over baseline.[182]

---

necropsy

Many types of studies require the postmortem examination of organs and tissues (necropsy). In addition, necropsy is frequently performed to diagnose disease problems.

➤ *Equipment*

Basic equipment needed to conduct a necropsy on a rabbit includes the following:

1. Latex or rubber gloves, lab coat, face mask, eye goggles or other protective eyewear.
2. A small metric ruler.
3. Toothed and serrated tissue forceps.
4. Scalpel blades and handles.
5. Dissecting and small operating scissors.
6. A probe.
7. Bone cutting forceps.
8. Sterile swabs for bacteriological culture of tissues.
9. Syringes (1 and 10 ml) with both large bore (18 gauge) and small bore (25 gauge) hypodermic needles.
10. Saline for washing of structures and paper towels for absorption of blood and other fluids are useful. Additional equipment may be useful and can be added to this basic kit.

> **Note:** Cutting surfaces of instruments will likely need to be honed periodically to maintain a sharp edge.

The necropsy is best performed in a dedicated necropsy room and on a surface that will facilitate drainage of blood and fluids and that can be easily cleaned and sanitized. Stainless steel necropsy tables are optimal, and some are designed with down-draft air flow to draw infectious agents and noxious odors away from personnel. If that type of equipment is unavailable, an area that is isolated from other animals, personnel areas, and feed and bedding storage could be used, provided that the area can be appropriately cleaned and sanitized following each use.

Formaldehyde, which is commonly used in diluted form as a tissue fixative, can cause allergic reactions and irritation of surfaces lined by mucous membranes.[188,189] In addition, form-aldehyde is considered to be a human carcinogen.[190] For these reasons, steps to limit exposure of personnel to formaldehyde should be taken, including adequate ventilation of the necropsy and tissue processing areas.

Personnel conducting necropsies should wear a clean lab coat, latex or rubber gloves, a face mask, and protective eyewear. Although specific pathogen-free rabbits harbor few infectious agents that would pose a risk to humans, this equipment will further decrease exposure of personnel to airborne allergens and formaldehyde, as well as protect clothes from soiling with blood or other material.

➤ *Necropsy technique*

Ideally, the rabbit should be necropsied immediately after death. Alternatively, carcasses may be stored for a short time (several hours) under refrigeration to delay tissue decomposition. Carcasses thus stored should be kept in refigerators not used for storage of food for animals or personnel. Freezing of carcasses can significantly interfere with meaningful necropsy. An in-depth description of necropsy methods for the rabbit can be found elsewhere.[191] General procedures for necropsy of a rabbit are as follows:

1. The rabbit is first examined externally for abnormalities such as discoloration, hair loss, wounds, masses, nasal or ocular discharge, and fecal or urine staining of the perineum. In addition, the oral cavity is examined, with particular attention paid to the teeth.

2. The skin is incised along the ventral midline with the scalpel blade, beginning at the lower jaw and continuing along the midline caudally to the pubis.

FIG. 47. Incised abdomen with exposed viscera during necropsy of a rabbit.

3. Using the scalpel, the skin is then gently reflected laterally and the subcutaneous tissues and underlying musculature are examined.

4. The abdominal wall is then incised and the abdominal cavity exposed using the dissecting scissors (Figure 47).

5. The organs and peritoneal surfaces are examined for abnormal coloration, size, presence of masses, traumatic damage, or any other abormal appearance. Depending on the time between death and necropsy and the carcass storage conditions, the tissues may appear abnormal due to postmortem autolysis, a natural process involving degradation of

tissues after death and unrelated to disease processes.

6. The thoracic cavity is exposed by cutting the diaphragm and then clipping the ribs using the bone cutting forceps. The clipped portion of the rib cage is then lifted off and removed or reflected laterally.

7. The lungs, heart, and pleural surfaces are examined for abnormalities as for the abdominal cavity. The organs are removed for inspection by cutting the trachea and cutting all attachments of trachea, lungs, and heart caudally to the diaphragm.

8. Abnormal fluids should be sampled for cytology and bacterial culture, and the volume and appearance of such fluids noted and recorded.

9. Other masses or abnormal tissues can be cultured using bacteriological culture swabs if infection is suspected.

10. Samples of tissues can be preserved in 10% neutral buffered formalin and saved for later histopathologic processing and evaluation. Smaller samples allow quicker penetration of formalin and are therefore preferred.

notes

resources

To provide the user of this handbook with information regarding sources of information, rabbits, equipment, and materials, examples of vendors and organizations are included in this chapter. The lists are not exhaustive, nor do they imply endorsement of one vendor over others. Rather, they provide a starting point for developing one's own list of resources. Sources for equipment are provided in Table 15 with contact information provided at the end.

organizations

A number of professional organizations exist which can serve as initial contacts for obtaining information regarding specific professional issues related to the care and use of laboratory rabbits. Membership in these organizations should be considered, since it allows the laboratory animal science professional to stay abreast of regulatory issues, improved procedures for the use of animals, management issues, and animal health issues. Relevant organizations include:

American Association for Laboratory Animal Science (AALAS), 70 Timber Creek Drive, Cordova, TN 38018 (Tel: 901-754-8620). AALAS serves a diverse professional group, ranging from principal investigators to animal care technicians to veterinarians. The journals, *Laboratory Animal Science* and

Contemporary Topics in Laboratory Animal Science are both published by AALAS and serve to communicate relevant information. AALAS sponsors a program for certification of laboratory animal science professionals at three levels: assistant laboratory animal technician (ALAT), laboratory animal technician (LAT), and laboratory animal technologist (LATG). The Institute for Laboratory Animal Management (ILAM) is a program designed to provide state of the art training in laboratory animal facility management. In addition, the association sponsors an annual meeting and an electronic bulletin board (LABBS). Local groups have also organized into smaller branches.

The **Laboratory Animal Management Association (LAMA)** serves as a mechanism for information exchange between individuals charged with management responsibilities for laboratory animal facilities. In this regard, the association publishes the *LAMA Review* and sponsors periodic meetings. The contact for LAMA changes annually with the elected president. The current contact for LAMA may be obtained from AALAS.

The **American Society of Laboratory Animal Practitioners (ASLAP)** is an association of veterinarians engaged in some aspect of laboratory animal medicine. The society publishes a newsletter to foster communication between members. In addition, the group sponsors annual meetings, generally in conjunction with annual meetings of AALAS and the American Veterinary Medical Association (AVMA). The contact for ASLAP changes annually with the elected president. Current contact information may be obtained from AALAS.

The **American College of Laboratory Animal Medicine (ACLAM)** is an association of laboratory animal veterinarians founded to encourage education, training, and research in laboratory animal medicine. ACLAM is recognized as a specialty of veterinary medicine by the AVMA and board certifies veterinarians as Diplomates in laboratory animal medicine by means of examination, experience requirements, and publication requirements. The group sponsors the annual ACLAM Forum as well as sessions at the annual AALAS meeting. Contact is established through the Executive Director, who at the time of publication is Dr. Charles McPherson, 200 Summerwinds Drive, Cary, NC 27511.

The **International Council for Laboratory Animal Science (ICLAS)** was organized to promote and coordinate the development of laboratory animal science throughout the world. ICLAS sponsors international meetings every fourth year, with regional meetings being held on a more frequent basis. The organization is composed of national, scientific, and union members. At the time of publication, the contact for ICLAS is Prof. Osmo Hänninen, Secretary General, Dept. of Physiology, University of Kuopio, P.O. Box 1627, SF-70211, Kuopio, Finland.

The **Institute of Laboratory Animal Resources (ILAR)** functions under the auspices of the National Research Council to develop and make available scientific and technical information on laboratory animals and other biologic resources. A number of useful publications are available from ILAR, including the *Guide for the Care and Use of Laboratory Animals* and the *ILAR Journal*. Contact with ILAR can be established at 2101 Consitution Avenue, NW, Washington, D.C. 20418 (Tel: 202-334-2590).

The **Association for Assessment and Accreditation of Laboratory Animal Care International, Inc. (AAALAC International)** is a nonprofit organization which provides a mechanism for peer evaluation of laboratory animal care programs. AAALAC accreditation is widely accepted as strong evidence of a quality research animal care and use program. Contact with AAALAC may be made through the Executive Director at 11300 Rockville Pike, Suite 1211, Rockville, MD 20852-3035 (Tel: 301-231-5353).

publications

A number of published materials are valuable as additional reference materials, including both books and periodicals.

Books

The following books may be worthwhile sources of additional information:

1. *The Laboratory Rabbit*, edited by P.J. Manning, D.H. Ringler, and C.E. Newcomer, 1994. Academic Press, Inc., 525 B Street, Suite 1900, San Diego, CA 92101.

2. *The Biology and Medicine of Rabbits and Rodents*, by J.E. Harkness and J.E. Wagner, 1995. Williams & Wilkins, Baltimore, MD 21298-9724.

3. *Formulary for Laboratory Animals*, by C.T. Hawk and S.L. Leary, 1995. Iowa State University Press, Ames, IA 50014.

4. *Laboratory Animal Anesthesia*, by P.A. Flecknell, 1987. Academic Press, Inc., 525 B Street, Suite 1900, San Diego, CA 92101.

5. *Handbook of Veterinary Anesthesia*, by W.W. Muir, J.A.E. Hubbell, R.T. Skarda, and R.M. Bednarski 1995. C.V. Mosby Co., 11830 Westline Industrial Drive, St. Louis, MO 63146.

6. *Antibodies: A Laboratory Manual*, by E. Harlow and D. Lane, 1988. Cold Spring Harbor Laboratory, Box 100, Cold Spring Harbor, NY 11724.

7. *Necropsy Guide: Rodents and the Rabbit*, by D.B. Feldman and J.C. Seely, 1988. CRC Press, Inc., 2000 Corporate Blvd. N.W., Boca Raton, FL 33431.

Periodicals
The following periodicals are excellent sources of current relevant information:

1. *Laboratory Animal Science*. Published by the American Association for Laboratory Animal Science. For contact information, see above listing for AALAS.

2. *Contemporary Topics in Laboratory Animal Science*. Published by the American Association for Laboratory Animal Science. For contact information, see above listing for AALAS.

3. *Laboratory Animals*. Published by Royal Society of Medicine Press, 1 Wimpole Street, London W1M 8AE, UK.

4. *Lab Animal*. Published by Nature Publishing Co., 345 Park Avenue South, New York, 10010-1707.

5. *ILAR Journal*. Published by the Institute of Laboratory Animal Resources, National Research Council. For contact information, see above listing for ILAR.

electronic resources

Many online sources of information relevant to the care and use of laboratory animals, including rabbits, are available. These include:

1. *Laboratory Animal Bulletin Board System (LABBS)*. This source includes message conferences on a variety of relevant topics. The system originates from the American Association for Laboratory Animal Science (AALAS), and that organization can be contacted for connection information.

2. *Comparative Medicine Discussion List (COMPMED)*. An electronic mailing list available through the internet, COMPMED is a valuable means to quickly tap into the expertise of laboratory animal science professionals around the world. At the time of publication, those interested in using this resource should subscribe to **listserv@wuvmd.wustl.edu** and mail to **compmed@wuvmd.wustl.edu**.

3. *Network of Animal Health (NOAH)*. NOAH is a commercial online service sponsored by the American Veterinary Medical Association. A number of forums cover a variety of topics, some of which would be of interest to those charged with the care and use of laboratory rabbits. Additional information can be obtained from the American Veterinary Medical Association (1931 N. Meacham Rd., Suite 100, Schaumburg, IL; 1-800-248-2862; e-mail: **72662.3435@compuserve.com**).

rabbits

Rabbits may be obtained from vendors of varying size and quality. Purchase of only specific pathogen-free (SPF) rabbits is

strongly encouraged. Vendors should be asked to supply information regarding the health status of their rabbit herd for
consideration prior to purchase. Small local or regional vendors
frequently offer quality rabbits at reasonable prices. In addition,
large vendors are often good sources of high quality rabbits with
known health status. It is impractical to list all vendors here,
however the following are examples of vendors which supply
rabbits:

1. Ace Animals, Inc., P.O. Box 122, Boyertown, PA 19512
    (Tel: 610-367-6047).

2. Buckshire, Inc., P.O. Box 155, 2025 Ridge Rd., Perkasie,
    PA 18944 (Tel: 1-800-229-2825).

3. Hare-Marland/Marland Breeding Farms, Inc., P.O. Box
    X, Hewitt, NJ 07421 (Tel: 201-728-3745).

4. HRP, Inc., P.O. Box 7200, Denver, PA 17517 (Tel:
    1-800-345-4114).

5. Millbrook Farm Breeding Labs, Box 513, Amherst, MA
    01004 (Tel: 413-253-5083).

6. Myrtle's Rabbitry, Inc., 4678 Bethesda Rd., Thompson
    Station, TN 37179 (Tel: 1-800-424-9511).

7. Western Oregon Rabbit Company, P.O Box 653, Philo-
    math, OR 97370 (Tel: 541-929-2245).

feed

Although small, local suppliers often can provide high quality
rabbit feed to research facilities, feed is more frequently acquired
from larger vendors such as those listed below:

1. Bio-Serv, Inc., P.O. Box 450, 8th & Harrison Streets,
    Frenchtown, NJ 08825 (Tel: 1-800-473-2155).

2. Harlan Teklad, Inc., P.O. Box 44220, Madison, WI
    53744-4220 (Tel: 1-800-483-5523).

3. PMI/Purina Mills, Inc., 505 North 4th St., P.O. Box 548, Richmond, IN 47375 (Tel: 1-800-227-8941).

4. United States Biochemical Corp., P.O. Box 22400, Cleveland, OH 44122 (Tel: 1-800-321-9322).

equipment

Sanitation

Several sources of disinfectants and other sanitation supplies are listed below:

1. BioSentry, Inc., 1481 Rock Mountain Blvd., Stone Mountain, GA 30083-9986 (Tel: 1-800-788-4246).

2. Convatec/Calgon Vestal Contamination Control, P.O. Box 147, St. Louis, MO 63166-0147 (Tel: 1-800-325-0966).

3. Rochester Midland, Inc., 333 Hollenbeck St., P.O. Box 1515, Rochester, NY 14603-1515 (Tel: 1-800-836-1627).

Cages and Research and Veterinary Supplies

Several sources for pharmaceuticals, hypodermic needles, syringes, surgical equipment, bandages, and other related items are provided below. Pharmaceuticals should generally be ordered and used only under the direction of a licensed veterinarian. Cages should meet the size requirements as specified by relevant regulatory agencies. Stainless steel is preferable to galvanized steel.

Possible Sources of Cages and Research and Veterinary Supplies

Item	Source
Cages and supplies	1,2,4,7,15,16,17,19
Veterinary and surgical supplies	8,12,13,25,26
Gas anesthesia equipment	12,19,24,25

Item (continued)	Source
Restrainers	1,4,5,11,12,14,15,20
Cervical collar	18, 25
Syringes and needles	5,8,11,13,26
Vascular access equipment	6,12,18,21
Osmotic pumps	3
Necropsy tools	5,11,25
Freund's adjuvant	9,11,23
Ribi adjuvant	22
Titermax adjuvant	10

contact information for cages and research and veterinary supplies

1. Allentown Caging Equipment, Inc., P.O. Box 698, Allentown, NJ 08501-0698 (Tel: 609-259-7951 or 1-800-762-2243).

2. Alternative Design Manufacturing and Supply, Inc., 16396 Highway 412, Siloam Springs, AR 72761 (Tel: 1-800-320-2459).

3. Alza Corporation, 950 Page Mill Road, P.O. Box 10950, Palo Alto, CA 94303-0802 (Tel: 1-800-692-2990).

4. Ancare Corp., 2475 Charles Court, P.O. Box 661, North Bellmore, NY 11710 (Tel: 1-800-645-6379).

5. Baxter Diagnostics, Inc., Scientific Products Divison, 1430 Waukegan Road, McGaw Park, IL 60085-9988 (Tel: 1-800-964-5227).

6. Braintree Scientific, Inc., P.O. Box 361, Braintree, MA 02184 (Tel: 617-843-2202).

7. Britz-Heidbrink, Inc., P.O. Box 1179, Wheatland, WY 82201-1179 (Tel: 307-322-4040).

8. Butler Co., Inc., 5000 Bradenton Ave., Dublin, OH 43017 (Tel: 1-800-225-7911).

9. Calbiochem-Novabiochem International, 10394 Pacific Center Court, San Diego, CA 92121 (Tel: 1-800-854-3417).

10. CytRx, Inc., 150 Technology Parkway, Technology Park/Atlanta, Norcross, GA 30092 (Tel: 404-368-9500).

11. Fisher Scientific, Inc., 711 Forbes Ave., Pittsburgh, PA 15219-4785 (Tel: 1-800-766-7000).

12. Harvard Apparatus, 22 Pleasant St., South Natick, MA 01760 (Tel: 1-800-272-2775).

13. IDE Interstate, Inc., 1500 New Horizons Blvd., Amityville, NY 11701 (Tel: 1-800-666-8100).

14. K.L.A.S.S., Inc., 4960 Aladen Exp., Suite 233, San Jose, CA 95118 (Tel: 408-266-1235).

15. Lab Products, Inc., 255 West Spring Valley Ave., P.O. Box 808, Maywood, NJ 07607 (Tel: 201-843-4600 or 1-800-526-0469).

16. Lenderking Caging Products, Inc., 1000 South Linwood Ave., Baltimore, MD 21224 (Tel: 410-276-2237).

17. Lock Solutions, Inc., P.O. Box 611, Kenilworth, NJ 07033 (Tel: 1-800-947-0304).

18. Lomir Biomedical, Inc., 99 East Main St., Malone, NY 12953 (Tel: 518-483-7697).

19. Otto Environmental, 6914 N. 124th St., Milwaukee, WI 53224 (Tel.: 1-800-484-5363 Ext. 6886).

20. P.A.M., Inc., 47 N. Front St., Souderton, PA 18964 (Tel: 215-723-0976 or 1-800-237-3373).

21. Pharmacia Deltec, Inc., St. Paul, MN 55112 (Tel: 1-800-433-5832).

22. Ribi Immunochem Research, Inc., P.O. Box 1409, Hamilton, MT 59840 (Tel: 406-363-6214 or 1-800-548-7424).

23. Sigma Chemical Co., Inc., P.O. Box 14508, St. Louis, MO (Tel: 314-771-5765 or 1-800-325-8070).

24. Vetamac, Inc., P.O. Box 178, Rossville, IN 46065 (Tel: 1-800-334-1583.

25. Viking Products, Inc., P.O. Box 2142, Medford Lakes, NJ 08055 (Tel: 609-953-0138).

26. J.A. Webster, Inc., 86 Leominster Road, Sterling, MA 01564 (Tel: 1-800-225-7911).

bibliography

1. Graur, D., Duret, L., and Gouy, M., Phylogenetic position of the order Lagomorpha (rabbits, hares and allies), *Nature*, 379, 333, 1996.

2. Jilge, B., The rabbit: a diurnal or nocturnal animal?, *J. Exp. Anim. Sci.*, 34, 170, 1991.

3. Harkness, J. E. and Wagner, J. E., *The Biology and Medicine of Rabbits and Rodents*, Williams & Wilkins, Baltimore, 1995.

4. Gentz, E. J., Harrenstein, L. A., and Carpenter, J. W., Dealing with gastrointestinal, genitourinary, and musculoskeletal problems in rabbits, *Vet. Med.*, 90, 365, 1995.

5. Jones, T. C. and Hunt, R. D., *Veterinary Pathology*, 5th edition, Lea & Febiger, Philadelphia, 1983, 1510.

6. Poole, T., *The UFAW Handbook on the Care and Management of Laboratory Animals*, 6th edition, Longman Scientific and Technical, Essex, 1987, 418.

7. Gillett, C. S., Selected drug dosages and clinical reference data, in *The Biology of the Laboratory Rabbit*, 2nd edition, Manning, P. J., Ringler, D. H., and Newcomer, C. E., Eds., Academic Press, San Diego, 1994, Appendix.

8. Kraus, A. L., Weisbroth, S. H., Flatt, R. E., and Brewer, N., Biology and diseases of rabbits, in *Laboratory Animal Medicine*, Fox, J. G., Cohen, B. J., and Loew, F. M., Eds., Academic Press, Orlando, 1984, chap. 8.

9. Kaplan, H. M. and Timmons, E. H., *The Rabbit — A Model for the Principles of Mammalian Physiology and Surgery*, Academic Press, New York, 1979.

10. Burns, K. F. and De Lannoy, C. W., Jr., Compendium of normal blood values of laboratory animals, with indications of variations, *Toxicol. Appl. Pharmacol.*, 8, 429, 1966.

11. Fox, R. R., The rabbit, in *The Clinical Chemistry of Laboratory Animals*, Loeb, W. F. and Quimby, F. W., Eds., Pergamon, New York, 1989.

12. Woolford, S. T., Schroer, R. A., Gohs, F. X., Gallo, P. P., Brodeck, M., Falk, H. B., and Ruhren, R., Reference range data base for serum chemistry and hematology values in laboratory animals, *J. Toxicol. Environ. Health*, 18, 161, 1986.

13. Loeb, W. F. and Quimby, F. W., *The Clinical Chemistry of Laboratory Animals*, Pergamon, New York, 1989, Appendix.

14. Curiel, T. J., Perfect, J. R., and Durack, D. T., Leukocyte subpopulations in cerebrospinal fluid of normal rabbits, *Lab. Anim. Sci.*, 32, 622, 1982.

15. Zurovsky, Y., Mitchell, G., and Hattingh, J., Composition and viscosity of interstitial fluid of rabbits, *Exp. Physiol.*, 80, 203, 1995.

16. Barzago, M. M., Bortolotti, A., Omarini, D., Aramayona, J. J., and Bonati, M., Monitoring of blood gas parameters and acid-base balance of pregnant and non-pregnant rabbits (*Oryctolagus cuniculus*) in routine experimental conditions, *Lab. Anim.*, 26, 73, 1992.

17. Sanford, T. D. and Colby, E. D., Effect of xylazine and ketamine on blood pressure, heart rate and respiratory rate in rabbits, *Lab. Anim. Sci.*, 30, 519, 1980.

18. Kabata, J., Gratwohl, A., Tichelli, A., John, L., and Speck, B., Hematologic values of New Zealand White rabbits determined by automated flow cytometry, *Lab. Anim. Sci.*, 41, 613, 1991.

19. Hafez, E. S. E., Rabbits, in *Reproduction and Breeding Techniques for Laboratory Animals*, Hafez, E. S. E., Ed., Lea & Febiger, Philadelphia, 1970, chap. 16.

20. Kennelly, J. J. and Foote, R. H., Superovulatory response of pre- and post-pubertal rabbits to commercially available gonadotropins, *J. Reprod. Fertil.*, 9, 177, 1965.

21. Williams, J. Gladen, B. C., and Turner, T. W., The effects of ethylene dibromide on semen quality and fertility in the rabbit: evaluation of a model for human seminal characteristics, *Fundam. Appl. Toxicol.*, 16, 687, 1991.

22. Foote, R. H. and Simkin, M. E., Use of gonadotropic releasing hormone for ovulating the rabbit model, *Lab. Anim. Sci.*, 43, 383, 1993.

23. DeBoer, K. F. and Krueger, D., A simplified artificial vagina for use in rabbits and other species, *Lab. Anim. Sci.*, 41, 187, 1991.

24. Fayrer-Hosken, R. A., Reynolds, D. C., and Brackett, B. G., An efficient rabbit artificial vagina and its use in assessing sperm fertilizing ability *in vitro*, *Lab. Anim. Sci.*, 37, 359, 1987.

25. Patton, N. M., Colony husbandry, in *The Biology of The Laboratory Rabbit*, Manning, P. J., Ringler, D. H., and Newcomer, C. E., Eds., Academic Press, San Diego, 1994, chap. 2.

26. Joint Working Group on Refinement, Refinements in rabbit husbandry, *Lab. Anim.*, 27, 301, 1993.

27. Committee to Revise the Guide for the Care and Use of Laboratory Animals, *Guide for the Care and Use of Laboratory Animals*, National Research Council, National Academy Press, Washington, D.C., 1996.

28. CFR (Code of Federal Regulations), Title 9; Parts 1, 2, and 3 (Docket 89-130), Federal Register, Vol. 54, No. 168, August 31, 1989, and 9 CFR Part 3, (Docket No. 90-218), Federal Register, Vol. 56, No. 32, February 15, 1991.

29. Hagen, K. W., Colony husbandry, in *The Biology of the Laboratory Rabbit*, Weisbroth, S. H., Flatt, R. E., and Kraus, A. L., Eds., Academic, Orlando, 1974, chap. 2.

30. O'Steen, W. K. and Anderson, K. V., Photically evoked responses in the visual system of rats exposed to continuous light, *Exp. Neurol.*, 30, 525, 1971.

31. O'Steen, W. K., and Anderson, K. V., Photoreceptor degeneration after exposure of rats to incandescent illumination, *Z. Zellforsch. Mikrosk. Anat.*, 127, 306, 1972.

32. Patton, N. M., Holmes, H. T., Caveny, D. D., Matsumoto, M., and Cheeke, P. R., Experimental inducement of snuffles in rabbits, *J. Appl. Rabbit Res.*, 3, 8, 1980.

33. Nayfield, K. C. and Besch, E. L., Comparative responses of rabbits and rats to elevated noise, *Lab. Anim. Sci.*, 31, 386, 1981.

34. Love, J. A., Group housing: meeting the physical and social needs of the laboratory rabbit, *Lab. Anim. Sci.*, 44, 5, 1994.

35. Whary, M., Peper, R., Borkowski, G., Lawrence, W., and Ferguson, F., The effects of group housing on the research use of the laboratory rabbit, *Lab. Anim.*, 27, 330, 1993.

36. Huls, W. L., Brooks, D. L., and Bean-Knudsen, D., Response of adult New Zealand White rabbits to enrichment objects and paired housing, *Lab. Anim. Sci.*, 41, 609, 1991.

37. Podberscek, A. L., Blackshaw, J. K., and Beattie, A. W., The behaviour of group penned and individually caged laboratory rabbits, *Appl. Anim. Behav. Sci.*, 28, 353, 1991.

38. Cheeke, P. R., Nutrition and nutritional diseases, in *The Biology of the Laboratory Rabbit*, Manning, P. J., Ringler, D. H., and Newcomer, C. E., Eds., Academic, San Diego, 1994, chap. 14.

39. Board on Agriculture and Renewable Resources, Subcommittee on Rabbit Nutrition, Committee on Animal Nutrition, *Nutrient Requirements of Rabbits*, 2nd rev. ed., National Academy of Sciences, Washington, D.C., 1977.

40. Slade, L. M. and Hintz, H. F., Comparison of digestion in horses, ponies, rabbits, and guinea pigs, *J. Anim. Sci.*, 28, 842, 1969.

41. Cheeke, P. R., Nutrition of the domestic rabbit, *Lab. Anim. Sci.*, 26, 654, 1976.

42. Chapin, R. E. and Smith, S. E., Calcium requirement of growing rabbits, *J. Anim. Sci.*, 26, 67, 1967.

43. Stevenson, R. G., Palmer, N. C., and Finley, G. G., Hypervitaminosis D in rabbits, *Can. Vet. J.*, 17, 54, 1976.

44. Ringler, D. H. and Abrams, G. D., Nutritional muscular dystrophy and neonatal mortality in a rabbit breeding colony, *J. Am. Vet. Med. Assoc.*, 157, 1928, 1970.

45. DiGiacomo, R. F., Deeb, B. J., and Anderson, R. J., Hypervitaminosis A and reproductive disorders in rabbits, *Lab. Anim. Sci.*, 42, 250, 1992.

46. Bondi, A. and Sklan, D., Vitamin A and carotene in animal nutrition, *Prog. Food Nutr. Sci.*, 8, 165, 1984.

47. Ringler, D. H. and Abrams, G. D., Laboratory diagnosis of vitamin E deficiency in rabbits fed a faulty commercial ration, *Lab. Anim. Sci.*, 21, 383, 1971.

48. Yamini, B. and Stein, S., Abortion, stillbirth, neonatal death, and nutritional myodegeneration in a rabbit breeding colony, *J. Am. Vet. Med. Assoc.*, 194, 561, 1989.

49. Wardrip, C. L., Artwohl, J. E., and Bennett, B. T., A review of the role of temperature vs. time in an effective cage sanitization program, *Cont. Topics Lab. Anim. Sci.*, 33 (5), 66, 1994.

50. Ryan, L. J., Maina, C. V., Hopkins, R. E., and Carlow, C. K. S., Effectiveness of hand cleaning in sanitizing rabbit cages, *Cont. Topics Lab. Anim. Sci.*, 32 (6), 21, 1993.

51. Animal Welfare Act, United States P.L. 89-544, 1966; P.L. 91-579, 1970; P.L. 94-279, 1976; and P.L. 99-198, 1985 (The Food Security Act).

52. Health Research Extension Act, United States P.L. 99-158, 1985.

53. Public Health Service Policy on Humane Care and Use of Laboratory Animals, Office for Protection from Research Risks, U.S. Department of Health and Human Services, Washington, D.C., 1986.

54. Bowman, P. J., A flexible occupational health and safety program for laboratory animal care and use programs, *Cont. Topics Lab. Anim. Sci.*, 30 (6), 15, 1991.

55. Hunskaar, S. and Fosse, R. T., Allergy to laboratory animals, in *Handbook of Laboratory Animal Science*, Svendsen, P., and Hau, J., Eds., CRC Press, Boca Raton, FL, 1994, chap. 7.

56. Centers for Disease Control and National Institutes of Health (CDC/NIH), *Biosafety in Microbiological and Biomedical Laboratories*, HHS Pub. No. (NIH) 93-8395, CDC/NIH, Atlanta, 1993.

57. Toth, L. A. and January, B., Physiological stabilization of rabbits after shipping, *Lab. Anim. Sci.*, 40, 384, 1990.

58. Scharf, R. A., Monteleone, S. A., and Stark, D. M., A modified barrier system for maintenance of *Pasteurella*-free rabbits, *Lab. Anim. Sci.*, 31, 513, 1981.

59. Carpenter, J. W., Mashima, T. Y., Gentz, E. J., and Harrenstein, L., Caring for rabbits: an overview and formulary, *Vet. Med.*, 90, 340, 1995.

60. Harrenstein, L., Gentz, E. J., and Carpenter, J. W., How to handle respiratory, ophthalmic, neurologic, and dermatologic problems in rabbits, *Vet. Med.*, 90, 373, 1995.

61. Pakes, S. P. and Gerrity, L. W., Protozoal diseases, in *The Biology of the Laboratory Rabbit*, Manning, P. J., Ringler, D. H., and Newcomer, C. E., Eds., Academic Press, San Diego, 1994, chap. 10.

62. Boriello, S. P. and Carman, R. J., Association of iota-like toxins and *Clostridium spiroforme* with both spontaneous and antibiotic-associated diarrhea and colitis in rabbits, *J. Clin. Microbiol.*, 17, 414, 1983.

63. Yonushonis, W. P., Roy, M. J., Carman, R. J., and Sims, R. E., Diagnosis of spontaneous *Clostridium spiroforme* iota enterotoxemia in a barrier rabbit breeding colony, *Lab. Anim. Sci.*, 37, 69, 1987.

64. Lipman, N. S., Weischedel, A. K., Connors, M. J., Olsen, D. A., and Taylor, N. S., Utilization of cholestyramine resin as a preventive treatment for antibiotic (clindamycin) induced enterotoxaemia in the rabbit, *Lab. Anim.*, 26, 1, 1992.

65. Gaertner, D. J., Comparison of penicillin and gentamicin for treatment of Pasteurellosis in rabbits, *Lab. Anim. Sci.*, 41, 78, 1991.

66. Okerman, L., Devriese, L. A., Gevaert, D., Uytterbroek, E., and Haesebrouck, F., *In vivo* activity of orally administered antibiotics and chemotherapeutics against acute septicaemic Pasteurellosis in rabbits, *Lab. Anim.*, 24, 341, 1990.

67. Broome, R. L. and Brooks, D. L., Efficacy of enrofloxacin in the treatment of respiratory Pasteurellosis in rabbits, *Lab. Anim. Sci.*, 41, 572, 1991.

68. Mahler, M., Stunkel, S., Ziegowski, C., and Kunstyr, I., Inefficacy of enrofloxacin in the elimination of *Pasteurella multocida* in rabbits, *Lab. Anim.*, 29, 192, 1995.

69. Suckow, M. A., Martin, B. J., Bowersock, T. L., and Douglas, F. A., Derivation of *Pasteurella multocida*-free rabbit litters by enrofloxacin treatment, *Vet. Microbiol.*, 51, 161, 1996.

70. Aramayona, J. J., Garcia, M. A., Fraile, L. J., Abadia, A. R., and Bregante, M. A., Placental transfer of enrofloxacin and ciprofloxacin in rabbits, *Am. J. Vet. Res.*, 55, 1313, 1994.

71. Aramayona, J. J., Mora, J., Fraile, L. J., Garcia, M. A., Abadia, A. R., and Bregante, M. A., Penetration of enrofloxacin and ciprofloxacin into breast milk, and pharmacokinetics of the drugs in lactating rabbits and neonatal offspring, *Am. J. Vet. Res.*, 57, 547, 1996.

72. Walker, R. C. and Wright, A. J., The quinolones, *Mayo Clin. Proc.*, 62, 1007, 1987.

73. Stahlmann, R., Cartilage damaging effect of quinolones, *Infection*, 19, S38, 1991.

74. DiGiacomo, R. F., Eradication programs for Pasteurellosis in rabbitries, *Cont. Topics Lab. Anim. Sci.*, 33, 69, 1994.

75. Cunliffe-Beamer, T. L. and Fox, R. R., Venereal spirochetosis of rabbits: eradication, *Lab. Anim. Sci.*, 31 (5), 379, 1981.

76. Vogtsberger, L. M., Harroff, H. Hugh, Pierce, G. E., and Wilkinson, G. E., Spontaneous dermatophytosis due to *Microsporum canis* in rabbits, *Lab. Anim. Sci.*, 36, 294, 1986.

77. Hagen, K. W., Ringworm in domestic rabbits: oral treatment with griseofulvin, *Lab. Anim. Care*, 19, 635, 1969.

78. Franklin, C. L., Gibson, S. V., Caffrey, C. J., Wagner, J. E., and Steffen, E. K., Treatment of *Trichophyton mentagrophytes* infection in rabbits, *J. Am. Vet. Med. Assoc.*, 198, 1625, 1991.

79. Curtis, S. K., Housley, R., and Brooks, D. L., Use of ivermectin for treatment of ear mite infestation in rabbits, *J. Am. Vet. Med. Assoc.*, 196, 1139, 1990.

80. Curtis, S. K. and Brooks, D. L., Eradication of ear mites from naturally infested conventional research rabbits using ivermectin, *Lab. Anim. Sci.*, 40, 406, 1990.

81. Bowman, D. D., Fogelson, M. L., and Carbone, L. G., Effect of ivermectin on the control of ear mites (*Psoroptes cuniculi*) in naturally infested rabbits, *Am. J. Vet. Res.*, 53, 105, 1992.

82. Leary, S. L., Manning, P. J., and Anderson, L. C., Experimental and naturally-occurring gastric foreign bodies in laboratory rabbits, *Lab. Anim. Sci.*, 34, 58, 1984.

83. Wagner, J. L., Hackel, D. B., and Samsell, A. G., Spontaneous deaths in rabbits resulting from gastric trichobezoars, *Lab. Anim. Sci.*, 24, 826, 1974.

84. Lee, K. J., Johnson, W. D., and Lang, C. M., Acute peritonitis in the rabbit (*Oryctolagus cuniculus*) resulting from a gastric trichobezoar, *Lab. Anim. Sci.*, 28, 202, 1978.

85. Gillett, N. A., Brooks, D. L., and Tillman, P. C., Medical and surgical management of gastric obstruction from a hairball in the rabbit, *J. Am. Vet. Med. Assoc.*, 183, 1176, 1983.

86. Bergdall, V. K. and Dysko, R. C., Metabolic, traumatic, mycotic, and miscellaneous diseases, in *The Biology of The Laboratory Rabbit*, Manning, P. J., Ringler, D. H., and Newcomer, C. E., Eds., Academic, San Diego, 1994, chap. 15.

87. Lindsey, J. R. and Fox, R. R., Inherited diseases and variations, in *The Biology of The Laboratory Rabbit*, Manning, P. J., Ringler, D. H., and Newcomer, C. E., Eds., Academic Press, San Diego, 1994, chap. 13.

88. Burrows, A. M., Smith, T. D., Atkinson, C. S., Mooney, M. P., Hiles, D. A., and Losken, H. W., Development of ocular hypertension in congenitally buphthalmic rabbits, *Lab. Anim. Sci.*, 45, 443, 1995.

89. Flecknell, P. A., *Laboratory Animal Anesthesia*, Academic Press, London, 1987, 129.

90. Bauck, L., Opthalmic conditions in pet rabbits and rodents, *Compend. Contin. Educ. Pract. Vet.*, 11, 258, 1989.

91. Latt, R. H., Drug dosages for laboratory animals, in *CRC Handbook of Laboratory Animal Science*, Vol. 3, Melby, E. C. and Altman, N. H., Eds., CRC Press, Boca Raton, 1976, 567.

92. Hawk, C. T. and Leary, S. L., *Formulary for Laboratory Animals*, Iowa State University Press, Ames, 1995, 43.

93. Wixson, S. K., Anesthesia and analgesia, in *The Biology of The Laboratory Rabbit*, Manning, P. J., Ringler, D. H., and Newcomer, C. E., Eds., Academic Press, San Diego, 1994, chap. 6.

94. Flecknell, P. A., *Laboratory Animal Anesthesia*, Academic Press, London, 1987, 99.

95. Bivin, W. S. and Timmons, E. H., Basic biomethodology, in *The Biology of The Laboratory Rabbit*, Weisbroth, S. H., Flatt, R. E., and Kraus, A. L., Eds., Academic Press, Orlando, 1974, p. 79.

96. Lipman, N. S., Marini, R. P., and Erdman, S. E., A comparison of ketamine/xylazine and ketamine/xylazine/acepromazine anesthesia in the rabbit, *Lab. Anim. Sci.*, 40, 395, 1990.

97. Poplilskis, S. J., Oz, M. C., Gorman, P., Florestal, A., and Kohn, D. F., Comparison of xylazine with tiletamine-zolazepam (Telazol) and xylazine-ketamine anesthesia in rabbits, *Lab. Anim. Sci.*, 41, 51, 1991.

98. Marini, R. P., Avison, D. L., Corning, B. F., and Lipman, N. S., Ketamine/xylazine/butorphanol: a new anesthetic combination for rabbits, *Lab. Anim. Sci.*, 42, 57, 1991.

99. Blake, D. W., Jover, B., and McGrath, B. P., Haemodynamic and heart rate reflex responses to propofol in the rabbit, *Br. J. Anaesth.*, 61, 194, 1988.

100. Robertson, S. A. and Eberhart, S., Efficacy of the intranasal route for administration of anesthetic agents to adult rabbits, *Lab. Anim. Sci.*, 44, 159, 1994.

101. Sanford, T. D. and Colby, E. D., Effect of xylazine and ketamine on blood pressure, heart rate, and respiratory rate in rabbits, *Lab. Anim. Sci.*, 30, 519, 1980.

102. Wyatt, J. D., Scott, R. A., and Richardson, M. E., The effects of prolonged ketamine-xylazine intravenous infusion on arterial blood pH, blood gases, mean arterial pressure, heart and respiratory rates, rectal temperature and reflexes in the rabbit, *Lab. Anim. Sci.*, 39, 411, 1989.

103. Beyers, T. M., Richardson, J. A., and Prince, M. D., Axonal degeneration and self-mutilation as a complication of the intramuscular use of ketamine and xylazine in rabbits, *Lab. Anim. Sci.*, 41, 519, 1991.

104. Lipman, N. S., Phillips, P. A., and Newcomer, C. E., Reversal of ketamine/xylazine anesthesia in the rabbit with yohimbine, *Lab. Anim. Sci.*, 37, 474, 1987.

105. Palmore, W. P., A fatal response to xylazine and ketamine in a group of rabbits, *Vet. Res. Commun.*, 14, 91, 1990.

106. Brammer, D. W., Doerning, B. J., Chrisp, C. E., and Rush, H. G., Anesthetic and nephrotoxic effects of Telazol in New Zealand White rabbits, *Lab. Anim. Sci.*, 41, 432, 1991.

107. Doerning, B. J., Brammer, D. W., Chrisp, C. E., and Rush, H. G., Nephrotoxicity of tiletamine in New Zealand White rabbits, *Lab. Anim. Sci.*, 42, 267, 1992.

108. Aeschbacher, G., and Webb, A. I., Propofol in rabbits 1. Determination of an induction dose, *Lab. Anim. Sci.*, 43, 324, 1993.

109. Aeschbacher, G. and Webb, A. I., Propofol in rabbits. 2. Long-term anesthesia, *Lab. Anim. Sci.*, 43, 328, 1993.

110. Flecknell, P. A., Anaesthesia, in *Laboratory Animal Anaesthesia*, Academic Press, London, 1987, chap. 3.

111. Muir, W. W. and Hubbell, J. A. E., Inhalation anesthesia, in *Handbook of Veterinary Anesthesia*, C.V. Mosby, St. Louis, 1989, chap. 9.

112. Flecknell, P. A., Cruz, I. J., Liles, J. H., and Whelan, G., Induction of anaesthesia with halothane and isoflurane in the rabbit: a comparison of the use of a face-mask or an anaesthetic chamber, *Lab. Anim.*, 30, 67, 1996.

113. Jordan, T., Freezing endotracheal tubes, *Tech Talk*, 1 (2), 1, 1996.

114. Davies, A., Dallak, M., and Moores, C., Oral endotracheal intubation of rabbits (*Oryctolagus cuniculus*), *Lab. Anim.*, 30, 182, 1996.

115. Bechtold, S. V. and Abrutyn, D., An improved method of endotracheal intubation in rabbits, *Lab. Anim. Sci.*, 41, 630, 1991.

116. Alexander, D. J. and Clark, G. C., A simple method of oral endotracheal intubation in rabbits (*Oryctolagus cuniculus*), *Lab. Anim. Sci.*, 30, 871, 1980.

117. Fick, T. E. and Schalm, S. W., A simple technique for endotracheal intubation in rabbits, *Lab. Anim.*, 21, 265, 1987.

118. Kruger, J., Zeller, W., and Schottmann, E., A simplified procedure for endotracheal intubation in rabbits, *Lab. Anim.*, 28, 176, 1994.

119. O'Roark, T. S. and Wilson, R. P., Use of the BAAM Mark VI for blind oral intubation in the rabbit, *Cont. Topics Lab. Anim. Sci.*, 34 (5), 87, 1995.

120. Bertolet, R. D. and Hughes, H. C., Endotracheal intubation: an easy way to establish a patent airway in rabbits, *Lab. Anim. Sci.*, 30, 227, 1980.

121. Flecknell, P. A., Liles, J. H., and Williamson, H. A., The use of lignocaine-prilocaine local anaesthetic cream for pain-free venepuncture in laboratory animals, *Lab. Anim.*, 24, 142, 1990.

122. McCormick, M. J. and Ashworth, M. A., Acepromazine and methoxyflurane anesthesia of immature New Zealand White rabbits, *Lab. Anim. Sci.*, 21, 220, 1971.

123. Freeman, M. J., Bailey, S. P., and Hodesson, S., Premedication, tracheal intubation and methoxyflurane anesthesia in the rabbit, *Lab. Anim. Sci.*, 22, 576, 1972.

124. Green, C. J., Neuroleptanalgesic drug combinations in the anaesthetic management of small laboratory animals, *Lab. Anim.*, 9, 161, 1975.

125. Hawk, C. T. and Leary, S. L., *Formulary for Laboratory Animals*, Iowa State University Press, Ames, 1995, 7.

126. Rapson, W. S. and Jones, T. C., Restraint of rabbits by hypnosis, *Lab. Anim. Care*, 14, 131, 1964.

127. Danneman, P. J., White, W. J., Marshall, W. K., and Lang, C. M., An evaluation of analgesia associated with the immobility response in laboratory rabbits, *Lab. Anim. Sci.*, 38, 51, 1988.

128. Koch, K. L. and Dwyer, A., Effects of acetylsalicylic acid on electromechanical activity of *in vivo* rabbit ileum, *Dig. Dis. Sci.*, 33, 962, 1988.

129. Liles, J. H. and Flecknell, P. A., The use of non-steroidal anti-inflammatory drugs for the relief of pain in laboratory rodents and rabbits, *Lab. Anim.*, 26, 241, 1992.

130. Marangos, M. N., Onyeji, C. O., Nicolau, D. P., and Night-ingale, C. H., Disposition kinetics of aspirin in female New Zealand White rabbits, *Lab. Anim. Sci.*, 45, 67, 1995.

131. Flecknell, P. A., Post-operative analgesia in rabbits and rodents, *Lab Animal*, 20, 34, 1991.

132. Portnoy, L. G. and Hustead, D. R., Pharmacokinetics of butorphanol tartarate in rabbits, *Am. J. Vet. Res.*, 53, 541, 1992.

133. Olson, M. E., Vizzutti, D., Morck, D. W., and Cox, A. K., The parasympatholytic effects of atropine sulfate and gly-copyrrolate in rats and rabbits, *Can. J. Vet. Res.*, 57, 383, 1994.

134. Linn, J. M. and Liebenberg, S. P., *In vivo* detection of rabbit astropinesterase, *Lab. Anim. Sci.*, 29, 335, 1979.

135. McCurnin, D. M. and Jones, R. L, Principles of surgical asepsis, in *Textbook of Small Animal Surgery*, Vol. 1, 2nd ed., Slatter, D. H., Ed., W. B. Saunders, Philadelphia, 1993, chap. 10.

136. Berg, R. J., Sterilization, in *Textbook of Small Animal Surgery*, Vol. 1, 2nd ed., Slatter, D. H., Ed., W. B. Saunders, Philadephia, 1993, chap. 11.

137. Wagner, S. D., Preparation of the surgical team, in *Textbook of Small Animal Surgery*, Vol. 1, 2nd ed., Slatter, D. H., Ed., W. B. Saunders, Philadelphia, 1993, chap. 12.

138. Powers, D. L., Assessment and preparation of the surgical patient, in *Textbook of Small Animal Surgery*, Vol. 1, 2nd ed., Slatter, D. H., Ed., W. B. Saunders, Philadelphia, 1993, chap. 13.

139. Hobson, H. P., Surgical facilities and equipment, in *Textbook of Small Animal Surgery*, Vol. 1, 2nd ed., Slatter, D. H., Ed., W. B. Saunders, Philadelphia, 1993, chap. 14.

140. AVMA Panel on Euthanasia, 1993 Report of the AVMA panel on euthanasia, *J. Am. Vet. Med. Assoc.*, 202, 229, 1993.

141. Joint Working Group on Refinement, Removal of blood from laboratory mammals and birds, *Lab. Anim.*, 27, 1, 1993.

142. Lacy, M. J., Kent, C. R., and Voss, E. W., d-Limonene: an effective vasodilator for use in collecting rabbit blood, *Lab. Anim. Sci.*, 37, 485, 1987.

143. Rao, S. B., Citrus oil: an effective vasodilator in rabbit blood collection, *Arch. Toxicol.*, 63, 79, 1989.

144. Cranney, J. and Zajac, A., A method for jugular blood collection in rabbits, *Cont. Topics Lab. Anim. Sci.*, 32 (6), 6, 1993.

145. Hall, L. L., Delope, O. H., Roberts, A., and Smith, F. A., A procedure for chronic intravenous catheterization in the rabbit, *Lab. Anim. Sci.*, 24, 79, 1974.

146. Melich, D., A method for chronic intravenous infusion of the rabbit via marginal ear vein, *Lab. Anim. Sci.*, 40, 327, 1990.

147. Martin, F., R., Alguacil, L. F., and Alamo, C., A method for catheterizing rabbit vena cava via marginal ear vein, *Lab. Anim. Sci.*, 41, 493, 1991.

148. Wallace, J., Gwynne, B., Dodd, J. R., and Davidson, R., Repeated arteriopuncture in the rabbit: a safe and effective alternative to cardiac puncture, *Anim. Technol.*, 39, 119, 1988.

149. Conn, H., Langer, R., Continuous long-term intra-arterial infusion in the unrestrained rabbit, *Lab. Anim. Sci.*, 28, 598, 1978.

150. Dennis, M. B., Jones, D. R., and Tenover, F. C., Chlorine dioxide sterilization of implanted right atrial catheters in rabbits, *Lab. Anim. Sci.*, 39, 51, 1989.

151. Perry-Clark, L. M. and Meunier, L. D., Vascular access ports for chronic serial infusion and blood sampling in New Zealand White rabbits, *Lab. Anim. Sci.*, 41, 495, 1991.

152. Hamory, B. H., Nosocomial bloodstream and intravascular device-related infections, in *Prevention and Control of Nosocomial Infections*, Wenzel, R. P., Ed., Williams & Wilkins, Baltimore, 1987, 283–319.

153. Penner, J., St. Claire, M., Beckwith, C., Besselsen, D., Wright, J., Fish, R., and Friskey, S., *Pseudomonas* sp. bacteremia associated with chronic vascular catheterization in rabbits, *Lab. Anim. Sci.*, 44, 642, 1994.

154. DaRif, C. A. and Rush, H. G., Management of septicemia in rhesus monkeys with chronic indwelling venous catheters, *Lab. Anim. Sci.*, 33, 90, 1983.

155. Danneman, P. J., Griffith, J. W., Beyers, T. M., and Lang, C. M., Renal and vascular damage associated with indwelling vascular access devices, *Lab. Anim. Sci.*, 38, 511, 1988.

156. Suckow, M. A., personal communication, 1996.

157. Kusumi, R. K. and Plouffe, J. F., A safe and simple technique for obtaining cerebrospinal fluid from rabbits, *Lab. Anim. Sci.*, 29, 681, 1979.

158. Hughes, P. J., Doherty, M. M., and Charman, W. N., A rabbit model for the evaluation of epidurally administered local anaesthetic agents, *Anaesth. Intens. Care*, 2, 298, 1993.

159. Bozkurt, P., Tunali, Y., and Kaya, G., Depth of the rabbit epidural space, *Anaesth. Intens. Care*, 23, 119, 1995.

160. Tissot van Patot, M. C., Seim, H. B., and Tucker, A., Catheterization of the subarachnoid space in rabbits using a vascular access port, *J. Invest. Surg.*, 8, 371, 1995.

161. Vistelle, R., Jaussaud, R., Trenque, T., and Wiczewski, M., Rapid and simple cannulation technique for repeated sampling of cerebrospinal fluid in the conscious rabbit, *Lab. Anim. Sci.*, 44, 362, 1994.

162. Haslberger, A. G. and Gaab, M. R., A technique for repeated sampling of pure cerebrospinal fluid from the conscious rabbit, *Lab. Anim. Sci.*, 36, 181, 1986.

163. Vistelle, R., Wiczewski, M., and Aurousseau, M., Continuous sampling of cerebrospinal fluid in the conscious rabbit, *Brain Res. Bull.*, 22, 919, 1989.

164. Sundberg, R. D. and Hodgson, R. E., Aspiration of bone marrow in laboratory animals, *Blood*, 4, 557, 1949.

165. Horan, P. K., Muirhead, K. A., Gorton, S., and Irons, R. D., Aseptic aspiration of rabbit bone marrow and enrichment for cycling cells, *Lab. Anim. Sci.*, 30, 76, 1980.

166. Powsner, E. R. and Fly, M. N., Aseptic aspiration of bone marrow from the living rabbit, *J. Appl. Physiol.*, 17, 1021, 1962.

167. Yu, C., Fiordalisi, I., and Harris, G. D., A unique swivel-tether system for continuous i.v. infusions in freely mobile unrestrained rabbits, *Cont. Topics Lab. Anim. Sci.*, 34 (2), 61, 1995.

168. Harris, G. D., Fiordalisi, I., and Yu, C., Intracranial pressure (ICP) during repair of diabetic ketoacidemia (DKA) in a rabbit model, *Pediatr. Res.*, 35, 203A, 1994.

169. Stills, H. F., Polyclonal antibody production, in *The Biology of The Laboratory Rabbit*, Manning, P. J., Ringler, D. H., and Newcomer, C. E., Eds., Academic Press, San Diego, 1994, chap. 20.

170. Vanatta, P. M., Capsule administration in the rabbit using a modified syringe, *Lab. Anim. Sci.*, 37, 367, 1987.

171. Eng, L. A., Carrig, C. B., Cordle, C. T., and Metz, C. B., Delivery of substances to the rabbit gastrointestinal tract by oral dosing with gelatin capsules, *Lab. Anim. Sci.*, 37, 239, 1987.

172. Rogers, G., Taylor, C., Austin, J. C., and Rosen, C., A pharyngostomy technique for chronic oral dosing of rabbits, *Lab. Anim. Sci.*, 38, 619, 1988.

173. Harlow, E. and Lane, D., *Antibodies: A Laboratory Manual*, Cold Spring Harbor Laboratory, Cold Spring Harbor, New York, 1988, 100.

174. Broderson, J. R., A retrospective review of lesions associated with the use of Freund's adjuvant, *Lab. Anim. Sci.*, 39, 400, 1989.

175. Johnston, B. A., Eisen, H., and Fry, D., An evaluation of several adjuvant emulsion regimens for the production of polyclonal antisera in rabbits, *Lab. Anim. Sci.*, 41, 15, 1991.

176. Smith, D. E., O'Brien, M. E., Palmer, V. J., and Sadowski, J. A., The selection of an adjuvant emulsion for polyclonal antibody production using a low-molecular-weight antigen in rabbits, *Lab. Anim. Sci.*, 42, 599, 1992.

177. Glenny, A. T., Pope, C. G., Waddington, H., and Wallace, U., Immunological notes, *J. Pathol.*, 29, 31, 1926.

178. Harlow, E. and Lane, D., *Antibodies: A Laboratory Manual*, Cold Spring Laboratory, Cold Spring Harbor, New York, 1988, 99.

179. Hillam, R. P., Tengerdy, R. P., and Brown, G. L., Local antibody production against the murine toxin of *Yersinia pestis* in a golf ball-induced granuloma, *Infect. Immun.*, 10, 458, 1974.

180. Clemons, D. J., Besch-Williford, C., Steffen, E. K., Riley, L. K., and Moore, D. H., Evaluation of a subcutaneously implanted chamber for antibody production in rabbits, *Lab. Anim. Sci.*, 42, 307, 1992.

181. Ried, J. L., Walker-Simmons, M. K., Everard, J. D., and Diani, J., Production of polyclonal antibodies in rabbits is simplified using perforated plastic golf balls, *Biotechniques*, 12, 660, 1992.

182. Anderson, J. A. and Henck, J. W., Toxicity and safety testing, in *The Biology of The Laboratory Rabbit*, Manning, P. J., Ringler, D. H., and Newcomer, C. E., Eds., Academic Press, San Diego, 1994, chap. 21.

183. Olson, C. T., Evaluation of the dermal irritancy of chemicals, in *Dermal and Ocular Toxicology*, Hobson, D. W., Ed., CRC Press, Boca Raton, FL, 1991, chap. 4.

184. Draize, J. H., Dermal toxicity, in *Appraisal of the Safety of Chemicals in Foods, Drugs, and Cosmetics*, Association of Food and Drug Officials of the United States, Austin, 1959, 46.

185. Daston, G. P. and Freeberg, F. E., Ocular irritation testing, in *Dermal and Ocular Toxicology*, Hobson, D. W., Ed., CRC Press, Boca Raton, FL, 1991, chap. 16.

186. Hansen, E. and Meyer, O., Animal models in reproductive toxicology, in *Handbook of Laboratory Animal Science, Vol. 2: Animal Models*, Svendsen, P. and Hau, J., Eds., CRC Press, Boca Raton, FL, 1994, chap. 3.

187. Weary, M. E. and Wallin, R. F., The rabbit pyrogen test, *Lab. Anim. Sci.*, 23, 677, 1973.

188. Misiak, P. M. and Miceli, J. N., Toxic effects of formaldehyde, *Lab. Manage.*, 24, 63, 1968.

189. Greenblatt, M., Swenberg, J., and Kang, H., Facts about formaldehyde, *Pathologist*, September, 648, 1983.

190. Formaldehyde Panel: Report of the Federal Panel on Formaldehyde, National Toxicology Program, Research Triangle Park, NC, 1980.

191. Feldman, D. B. and Seely, J. C., *Necropsy Guide: Rodents and the Rabbit*, CRC Press, Boca Raton, FL, 1988, chap. 6.

index

drug dosages for treatment,
53
transmission to humans,
35
Dewlap, 3
Dexamethasone, dosage, 53
Diaphragm, 70
Diarrhea, 40, 41
as clinical sign, 42, 43
general treatment, 53–54
Diastolic pressure, arterial, 7
Diazepam, 64
Disease, 7
drug dosages for treatment,
52–53
prevention through sanitation,
55
Disinfectants, 23–24
suppliers, 113
for veterinary care, 37
Diurnal behavior, 2
Does, 1
breeding life, 8
number of nipples, 3
reproductive tract, 4
teratogenic testing and,
101
territoriality of, 9
urethral orifice of, 4
Doors
of animal room, 12
of cage, 14
Dosages
analgesics, 65
anesthestics, 57
sedatives and tranquilizers,
64
for treatment, 52–53
Doxapram, dosage, 53
Drug dosages for treatment.
See Dosages
Drug Enforcement Agency
registration, 70
Ear pinch test during anesthesia,
67
Ears, 3
blood sampling from the, 76–77
clinical signs of disease, 42

ectoparasitic otitis externa, 42,
47–48
intravascular administration,
87
of kits, days until open, 10
physical examination of, 38
placed over eyes during
hypnosis, 73
tattoo identification on the,
28
torticollis with infection of
inner, 44, 45
Ear tags for identification, 27,
28
Ectoparasitic otitis externa, 42,
47–48
Eimeria irresidua, 42
Eimeria magna, 42
Eimeria stiedae, 42
Electrocardiogram, 67
Electronic resources, 111
Elizabethan collar, 69
Embryo, 100
EMLA cream, 63
Employees. *See* Personnel
Endotracheal intubation, 59,
60–62
Enrofloxacin, 44, 53
Enterotoxemia, clostridial, 42,
43–44
Environmental conditions
in animal room, 15–16
during transportation, 27
Environmental control, 12
Environmental enrichment,
16–17
Eosinophilic granules, 7
Eosinophils, 9
EPA (U.S. Environmental
Protection Agency), 32
Equipment
in animal room, 12
for endotracheal intubation, 60
protective. *See* Protective
equipment
suppliers, 113–114
Estrous cycle, lack of distinct,
8

Ulcerative pododermatitis, 42, 47, 48
Ulcers, 42
Urethra, 80, 81
Urethral orifice, 4
Uric acid, 6
Urinary bladder, 49, 80, 81, 82, 83
Urine, 2, 4. *See also* Sanitation
contamination of feeder, avoidance of, 17
excretion of calcium via, 19
pH, 4, 5
removal of scale, 23
sampling techniques, 80–84
specific gravity, 5
volume, 5
Urogenital system, 4
U.S. Department of Agriculture (USDA), 31
U.S. Environmental Protection Agency (EPA), 32
U.S. Food and Drug Administration (FDA), 32
Uterine adenocarcinoma, 42, 46
Uterine horns, 4
Uterus, 101

Vagina, 4, 80, 83
Vascular access port, 88
Vascular catheterization, 79–80
Vasodilation, 76–77
Vendors, for rabbits, 111–112
Venereal spirochetosis, 42, 46, 53
Ventilation, 16
Ventricles
fourth, 84, 85
third, 86
Veterinary care, 37–70
clinical problems, 40–52

physical examination, 38–39, 40
quarantine, 39–40
recovery from anesthesia, 68
supplies, 37–38, 113–114
Viscosity of interstitial fluid, 7
Vitamin A, nutritional requirement for, 19
Vitamin B, soft feces as source of, 20
Vitamin D, 19
Vitamin E, nutritional requirement for, 19
Vomit, inability rabbits to, 66
Vulva, 82
clinical signs of disease in the, 42, 46
during receptivity, 9

Water
oral administration through, 94
temperature for cleaning, 24
during transportation, 26
Watering devices, 15, 20
Water intake requirements, 5, 20
Weight. *See* Body weight
White blood cells, 9
in cerebrospinal fluid, 6
Work records, 29

Xylazine, 56, 57, 58, 64, 65
Xylene, 77

Yohimbine, 56
Yolk sac, 4

Zolazepam, 57, 58